Ecology

for dummies®
A Wiley Brand

Ecology

by René Fester Kratz, PhD

for dummies®
A Wiley Brand

Ecology For Dummies®

Published by: **John Wiley & Sons, Inc.**, 111 River Street, Hoboken, NJ 07030-5774, www.wiley.com

For general information on our other products and services, please contact our Customer Care Department within the U.S. at 877-762-2974, outside the U.S. at 317-572-3993, or fax 317-572-4002. For technical support, please visit https://hub.wiley.com/community/support/dummies.

Wiley publishes in a variety of print and electronic formats and by print-on-demand. Some material included with standard print versions of this book may not be included in e-books or in print-on-demand. If this book refers to media that is not included in the version you purchased, you may download this material at http://booksupport.wiley.com. For more information about Wiley products, visit www.wiley.com.

Library of Congress Control Number: 2026946324

ISBN 978-1-394-41507-6 (pbk); ISBN 978-1-394-41509-0 (ebk); ISBN 978-1-394-41508-3 (ebk)

SKY10163709_072726

Contents at a Glance

Table of Contents

Introduction

Human beings are part of an interconnected web of life that weaves through — and is dependent upon — the physical environment of planet Earth. As organisms seek out what they need to survive, they impact each other and their environment. The rapid growth of the human population is causing unprecedented effects on the ecosystems that sustain us. Cultivating a deep understanding of these ecosystems through the study of ecology can help us figure out how to live more sustainably on our planet.

Many factors influence the number and diversity of organisms that can live in each of the distinct environments on Earth. Climate and geography drive the evolution of unique communities that are adapted to their specific environment. While the organisms found in any given community may be different, all communities have examples of organisms that perform similar roles: producers make food, consumers eat others, and decomposers recycle the dead. All you have to do is turn on a nature documentary to see all the amazing ways that organisms have adapted for survival and reproduction in their environments. Understanding the specifics of how these organisms interact with each other and the physical world to get what they need is at the heart of the study of ecology.

About This Book

Ecology For Dummies explores fundamental concepts of ecology that are essential to an understanding of the factors that shape the living world, including human impacts that threaten the health of ecosystems. My goal is to explain each topic clearly, keeping scientific jargon to a minimum. I want this book to be understandable to anyone who picks it up, even if they don't have a science background. Along the way, I share lots of analogies and examples to help you wrap your brain around the main concepts.

Within this book, you may note that some web addresses break across two lines of text. If you're reading this book in print and want to visit one of these web pages, simply type in the web address exactly as it's noted in the text, pretending as though the line break doesn't exist. If you're reading this in e-book form, you've got it easy — just click the web address to be taken directly to the web page.

Conventions Used in This Book

To help you navigate through this book, I use the following conventions:

- *Italic* is used to emphasize and highlight new words or defined terms.
- **Boldfaced** text indicates keywords in bulleted lists or the action part of numbered steps.
- Monofont is used for web addresses.
- Sidebars, which look like text enclosed in a shaded gray box, consist of information that's interesting to know but not necessarily critical to your understanding of the chapter or section topic.

Foolish Assumptions

As I wrote this book, I tried to imagine who you are and what you need to understand ecology. Here's what I came up with:

- You're a college student who is taking an ecology class to help fulfill your degree requirements. For you, the chapters in the book are organized in a way that should match your college class fairly closely. If you want help following along in class, try reading the relevant sections in this book before you go to a lecture on a particular topic. Or, if you want to reinforce a concept that you heard about in class, read the related section after class.
- You're someone who wants to know a little bit more about how living things interact with the environment, including climate. Maybe you've heard about global warming and possible extinctions, and you'd like to understand how and why things are changing. You can start reading whatever section of the book interests you the most. For you, I've included references to other chapters whenever two subjects connect so that you can navigate easily to related topics.

Icons Used in This Book

I use some of the familiar *For Dummies* icons to help guide you and give you new insights as you read the material. Here's the scoop on what each one means.

REMEMBER

I use this icon to emphasize main ideas that you should definitely keep in mind.

TECHNICAL STUFF

I use this icon to flag detailed information that isn't essential to the main concept or process being presented. If you're not a student in a college-level ecology class, you can definitely skip this material.

TIP

I use this icon to present study tips or other information that can help you navigate through difficult material.

WARNING

I use this icon to flag potentially confusing ideas or common wrong ideas that people typically have about how something works. I know about these danger spots from my years of teaching, and I've flagged them to help you avoid these pitfalls.

Beyond the Book

In addition to the abundance of information and guidance related to ecology that we provide in this book, you get access to even more help and information online at `Dummies.com`. Check out this book's online Cheat Sheet. Just go to `www.dummies.com` and search for "Ecology For Dummies Cheat Sheet."

Where to Go from Here

With *Ecology For Dummies*, you can start anywhere in the book that you want. If you want to know more about how living things interact with each other, start with Chapter 7. If you've heard a lot about climate change or global warming and want to understand it better, check out Chapter 12. If you're worried about endangered species and want to know more about why they're in trouble, Chapter 13 is for you. And if you want to know more about what ecologists are doing to try and maintain the health of ecosystems and save biodiversity, check out Chapter 14.

1 Responding to the Physical Environment

IN THIS PART . . .

Connect with the natural world

Discover how climate is fundamental

Evolve with the times

Learn to adapt

IN THIS CHAPTER

» **Considering the impact of ecology**

» **Interacting with others and the environment**

» **Using ecological knowledge to restore balance**

Chapter **1**

Making Connections with Ecology

Ecology is the science of studying the relationships between all living things and our home, planet Earth (*oikos* = house and family). The scope of ecology is as big as the planet itself, seeking to answer questions that focus on small details, like the role of a single species in an ecosystem, to those that look at the big picture, like how climate affects the distribution of living things across the planet's surface.

The science of ecology is making important contributions to the human worldview, helping people see that we are both embedded in and dependent upon a web of life. Figure 1-1 illustrates this ecological understanding and contrasts it with an older view that sees humans as superior to other species. At its extreme, the older worldview leads people to use the natural world to serve human needs without regard to possible impacts on other species. In contrast, an ecological understanding promotes awareness of the importance of maintaining ecosystem health for humans and other species. The human population is more than 8 billion and still growing, stretching resources thin and straining ecosystems all around the world. Ecology looks at how the pieces of an ecosystem fit together to make a functioning whole and how ecosystems can recover after disturbance. Ecology is needed now more than ever as people look for solutions to modern environmental problems.

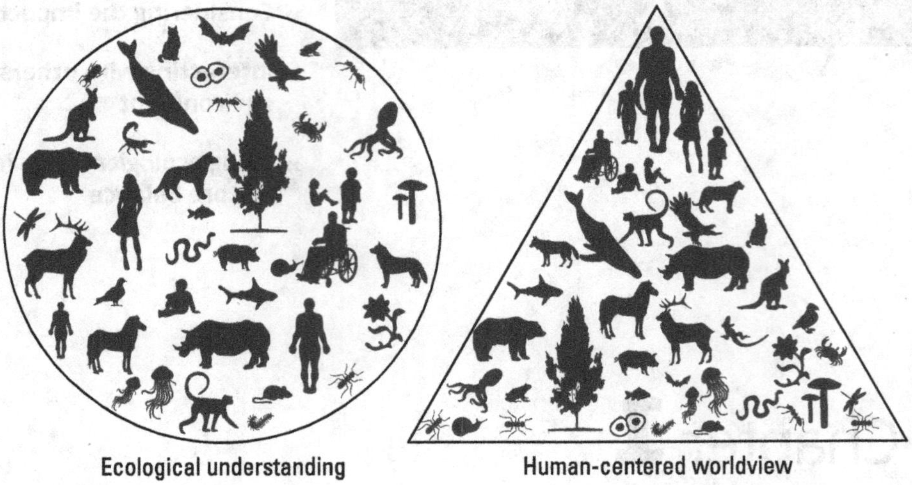

FIGURE 1-1: Ecological understanding.

Evolving to Fit the Physical World

The Earth is unique among all the known planets because it is covered with life.

REMEMBER

The *biosphere* is the interacting web of life that covers the planet's surface, extending into the Earth's crust, through the oceans, and up into the lower atmosphere.

Scientists think life is possible on Earth because of our position in the solar system and the physical properties of our planet. Earth orbits the sun at just the right distance to put it in the Goldilocks zone, neither too hot nor too cold. It has an atmosphere that protects the surface of the planet from harmful radiation and creates the pressure that makes liquid water possible. The Earth's core generates a massive magnetic field that helps protect the atmosphere from solar winds. These examples illustrate the deep connection between physical conditions and the biosphere.

Climate shapes our world

The physical conditions on planet Earth shape the environment and thus determine where organisms can live and grow.

REMEMBER

An *ecosystem* is a group of organisms interacting with each other and their environment.

The amount of solar radiation, combined with the movement of heat by wind and water currents, creates distinct climate zones on the planet (Chapter 2). Within each climate zone, differences in geology and topography further define unique environments that support distinct plant communities. These distinct large-scale communities are called *biomes*.

REMEMBER

Biomes are ecosystem types with similar climate, animals, and plants.

Ecologists generally agree that six major biomes exist on the planet: aquatic, forest, desert, tundra, grassland, and thick subarctic forests called taiga, or boreal forests. These biomes can be further divided into subcategories. Ecosystems from two completely different locations on the planet can be recognized as belonging to the same biome type. Although the species in each ecosystem may be unique, they often have similar structures and functions that enable them to live in that environment.

Living things are adaptable

Living things are found in almost all environments on the Earth, including the driest deserts, Antarctic ice, lightless caves, and deep ocean vents. In each of these environments, organisms have developed the necessary structures and strategies for success. In ecology and other biological sciences, an organism's success is ultimately defined as its *biological fitness*.

REMEMBER

Biological fitness is the ability of an organism to survive and reproduce, passing its genes to the next generation.

In any given environment, the individuals with the most advantageous traits are more likely to survive and reproduce. Scientists call this *natural selection* or *survival of the fittest* (Chapter 3). Over time, the most advantageous traits become more common in the population.

REMEMBER

Evolution by natural selection occurs when populations change over time to become more adapted to their environment.

The characteristics that help organisms succeed in a particular environment are called *adaptations* (Chapter 4). If someone asked you what traits organisms that live in the water needed to be successful, you might think about the streamlined bodies and fins of fish or the ability of whales to hold their breath for long periods of time. These are examples of adaptations to aquatic life. Likewise, organisms on land developed adaptations like the roots of plants for getting water or the wings of bats and birds for flying to escape predators. When ecologists look at unrelated species that evolved in similar biomes, they often find similar adaptations.

REMEMBER

Convergent evolution (Chapter 3) occurs when organisms that aren't closely related evolve similar characteristics, often as solutions to the same environmental problems (Chapter 4).

Reproducing into Populations

At the same time that the human population has grown to its largest size yet (more than eight billion people), thousands of other species have declined to the brink of extinction. Ecologists study the factors that affect populations, including how they grow, change, move, and decline, in order to predict how populations may change in the future. Ecologists can use these insights to develop effective strategies for the protection of threatened species and to manage populations of economic or ecological importance.

Living your life history

Natural selection favors traits that maximize reproductive success, but the road to success isn't the same for all species. Some species mature quickly and reproduce rapidly, producing vast numbers of offspring in the hopes that some will survive. Other species develop more slowly, growing and storing resources before they reproduce. Then, they produce just a few offspring at a time, which they tend carefully to increase the chance of survival. These two species could be equally successful, ultimately producing the same number of surviving offspring. The difference between the two is in the strategy they use to allocate their available resources.

REMEMBER

Life history strategies (Chapter 5) are the overall patterns and timing of events in the life history of a species, such as age of maturity, number of reproductive events, number of offspring produced, and lifespan.

As individuals reproduce, they produce groups of related organisms called populations.

REMEMBER

A *population* (Chapter 6) is a group of organisms that live in the same area at the same time and have the potential to interbreed.

Ecologists study populations to gain insight into the factors that shape their growth, distribution, and interactions. Population growth is affected by birth and death rates, as well as immigration and emigration. The balance of birth and death rates can affect the age structure of a population. For example, if the birth rate is very high, a population might have more young individuals than older ones. Populations may be distributed evenly over an area, or they may show a clumped distribution with many individuals in certain areas and fewer in others.

Interacting with others

Individuals in populations interact with each other and with members of populations of other species. Every possibility that you can imagine, from helpful to harmful, is a potential outcome of these interactions (Chapter 7). One of the most obvious interactions is a predator-prey relationship where one species eats another. A more subtle negative impact can occur when two individuals compete for a limited resource or if a parasite slowly drains resources from its host. Some of the most fascinating interactions are those that result from mutualism, where two species form a lifelong symbiotic partnership that benefits both parties. Each of the types of interactions impacts the success of the individuals involved and can also affect ecosystem function.

Forming Communities and Ecosystems

In nature, populations of different species interact with each other and their environment to form ecosystems.

Establishing a community

Populations of different species combine to form *communities* (Chapter 8). The function of the individuals within a community and their interactions with each other give communities distinct characteristics.

REMEMBER

Two important factors that influence *community structure* are the number of different species it contains and the identities of those organisms.

Some organisms have especially powerful effects on community structure:

- ***Dominant species* are the most common or largest species in a community.** They may affect community structure by outcompeting other species for resources or by altering environmental conditions such as nutrient and water availability. Humans are the dominant species in most parts of the world.
- ***Foundation species* create or maintain the physical or chemical environment that other organisms depend on.** Mangrove trees are a foundation species because their roots stabilize coastlines and provide breeding grounds for many species of fish. Coastal mangrove forests are home to insects and provide nesting grounds for shorebirds and migratory birds.

» ***Keystone species* keep a community in balance by controlling populations of other species.** Ochre seastars are keystone species in the rocky intertidal of the Pacific Northwest Coast. They prey on mussels, preventing the mussels from taking over rock surfaces. This allows other species to grow on the rocks, increasing the diversity of these intertidal communities.

Communities change over time in response to the activities of species within the community. For example, as seedlings grow into taller plants, they create shade that may favor the growth of some species but inhibit the growth of others.

REMEMBER

Ecological *succession* refers to the process by which communities and their environment change over time.

Over time, communities can change slowly by succession until they reach a stable *climax community* that is distinctive to a particular area. Examples of these late successional communities include the coastal redwood forests of Northern California or the Northern hardwood forests of the Eastern United States.

Most of the time, however, communities cycle between different phases of the succession process as they respond to disturbances (Chapter 9). Disturbances come in many forms, from biological sources such as grazing animals to destructive environmental hazards such as severe storms or forest fires. After a disturbance, *pioneer species* that have adaptations to survive in harsh environments move in. As they grow, they may change the environment in ways that facilitate the growth of other species. As these species grow, they may facilitate or inhibit others, and the community structure continues to change over time.

Getting what you need to survive

Living things all have a fundamental need for energy and matter. Energy allows organisms to move and grow, and matter is the building material that forms cells and bodies. Every ecosystem relies upon producers that capture energy from the environment and use it to build food, handy little packages of carbohydrates, proteins, or fats that contain both matter and energy.

For example, plants use *photosynthesis* (see Chapter 10) to capture energy from the sun, using it to build carbohydrates out of carbon dioxide and water. The atoms in carbon dioxide (CO_2) and water (H_2O) are the physical matter necessary to build carbohydrates like the simple sugar glucose ($C_6H_{12}O_6$). Once producers like plants make food, then other organisms in the ecosystem can eat them to get the food (so rude!), using it as their source of both matter and energy.

As living things transfer energy from food to do work in their cells, some of the energy transforms into heat that moves into the atmosphere. Heat isn't useful to producers as a source of energy for food, so ecosystems require a steady input of new energy, like the light plants receive from the sun.

REMEMBER

Energy constantly flows through ecosystems, moving from its initial capture by producers to other organisms, and then back out to the environment as heat (see Chapter 10).

Organisms break down food molecules to access the stored energy. When they use *cellular respiration* (see Chapter 10) to extract energy, atoms in the food get released back to the environment in the form of carbon dioxide and water. These molecules can get picked up again by plants and reused for photosynthesis, forming a loop which is part of the *carbon cycle* (Chapter 11). The carbon cycle describes how carbon atoms move through ecosystems, passed along from one organism to the next, and sometimes spend time in the environment.

The carbon cycle is just one example of the many ways that organisms interact to share and recycle matter. All the elements that organisms need to build their cells and function come from the environment, picked up by an organism that has the right skills, then passed around as organisms eat, breathe, and die.

REMEMBER

Scientists trace the movement of different types of atoms through their changing forms, drawing pathways called *biogeochemical cycles* (see Chapter 11), which show how matter cycles through ecosystems.

The Big Picture

The ecosystems on Earth have been changing for at least 3.5 billion years since the first cells evolved. Photosynthetic organisms created a dramatic change when they added oxygen to our atmosphere about 2.4 billion years ago. Around the same time, the first major ice age occurred, spreading ice sheets across the Earth all the way to the equator.

Since then, the ice has ebbed and flowed through seven more ice ages. In between the periods of glacial expansion, the Earth warmed, biomes shifted, and groups of species expanded and shrank. The age of reptiles brought the dinosaurs to prominence between 245 and 66 million years ago, after which the age of mammals began. Many scientists say that we're now living in a new age called the Anthropocene, which marks the dominance of humans.

Living in changing times

Although the Earth has a long record of change, it may never have changed as fast as it is today. For example, global temperatures are rising faster than they have in 10,000 years.

REMEMBER

Ninety-seven percent of scientists agree that the rapid changes in our climate happening today result from human actions.

During the Industrial Revolution, humans began to greatly increase the amount of fossil fuels they burned. The burning of fossil fuels releases greenhouse gases like carbon dioxide into our atmosphere.

REMEMBER

Greenhouse gases (Chapter 12) have a chemical structure that makes them excellent at absorbing heat.

The increase in greenhouse gases traps more heat in our atmosphere, which is causing the Earth's temperature to rise. The rising temperatures are having large-scale impacts on our climate because they're altering the flow of wind and water currents. Severe storms and wildfires are more common, resulting in increased disturbance to ecosystems. Great carbon dioxide concentrations in the ocean result in acidification, which kills the corals that are foundation species in coral reefs. Ice sheets are melting, raising water levels and affecting coastal areas.

Increased disturbance and global temperatures add to the pressures that put species at risk for extinction. Loss of species decreases diversity in ecosystems, reducing their resilience.

REMEMBER

Biodiversity (Chapter 13) is the variability in living organisms in an ecosystem.

In addition to climate change, five major factors are accelerating extinction rates. Ecologists use the acronym HIPPO to summarize these factors, which are discussed in Chapter 13:

- Habitat loss,
- Invasive species,
- Pollution,
- Population growth (human),
- Overharvesting.

From an ecological perspective, all living things are connected to each other and to our environment. When we have diversity, we have strength and balance. When one organism makes waste, another organism might be able to use it, changing it into something else. We have enough food and clean water. As we lose diversity, we begin to lose these things.

Saving what matters

As we live through this time of rapid change, we can use the lessons from ecology to try to make things better for other species and our own.

REMEMBER

Conservation ecology (Chapter 14) is the branch of ecology that seeks to preserve and maintain biodiversity and natural ecosystems.

Conservation ecologists seek to understand human impacts on ecosystems and develop ways to counteract those impacts. For example, habitat fragmentation is pushing species toward extinction. Fragmentation happens when patches of habitat are disturbed, converting a large habitat into a bunch of smaller patches. Fragmentation can have many negative impacts on wildlife, including:

- **Increased edge effects:** The edge of a habitat is different than the interior. It may receive more light and have less moisture, leading to the growth of different plant species. It may also be riskier because of exposure to predators.
- **Inability to access potential mates:** Species that can't seek mates in a wider area may start to inbreed, which can reduce genetic diversity and contribute to the risk of extinction.
- **Loss of habitat area:** Some species can only live in one particular habitat. As patches of this habitat are lost, the number of individuals in this species is reduced.

One strategy to combat these negative outcomes is to create wildlife corridors between habitat fragments. These corridors allow individuals to move safely among the fragments, increasing foraging and breeding areas.

Another solution to human disturbance is to allow ecosystems to recover by succession. Unfortunately, the rate of recovery of ecosystems after a disturbance is often slower than the rate of disturbance due to human activity. By studying natural succession, however, ecologists can identify ways to support and facilitate recovery.

REMEMBER

Restoration ecology (Chapter 14) uses ecological principles to restore degraded ecosystems to conditions as similar as possible to their natural state.

Some examples of ways that restoration ecologists facilitate recovery include removing the source of the disturbance, planting native species, and enhancing habitats for desired species.

In addition to restoring degraded ecosystems, restoration ecologists can help plan for the changing times ahead. Species are already moving in response to climate change as former habitats become unsuitable. To save some species, we may need to create habitats for species that are displaced.

IN THIS CHAPTER

» Getting energy from the Sun

» Moving energy around the Earth

» Touring Earth's biomes

Chapter 2
Establishing Climate

The physical environment varies over the surface of the Earth, from the warm, humid tropics at the equator to the cold, dry tundra at the poles. In the short term, physical conditions determine whether organisms can survive in a particular location. In the long term, these conditions shape the evolution of species, resulting in the amazing diversity of life on Earth today. In this chapter, you'll explore the factors that determine the physical environments that shape ecosystems.

Transmitting Energy to the Earth

The Sun provides a constant stream of energy to the Earth in the form of solar radiation. The energy passes through the atmosphere, arriving at the surface where it warms the land, water, and air. The warming of air and water causes them to move and circulate around the Earth, distributing energy and contributing to the formation of different climate zones.

REMEMBER

Climate refers to the long-term environmental conditions found in a specific area, including the timing and patterns of precipitation, temperature, and wind.

WARNING

Climate isn't the same thing as *weather*, which refers to the atmospheric conditions at any given moment. Although both can include descriptions of factors like precipitation, temperature, and humidity, weather refers to short-term conditions, while climate refers to their long-term averages and ranges.

Surface temperature is an important component of climate because it controls many environmental and ecological processes that are essential to humans and other life, including the water cycle (Chapter 4), the carbon cycle (Chapter 11), and the kinds of living things that can survive in different places on Earth.

REMEMBER

Temperature is a measure of the average kinetic energy of molecules in a substance. *Heat* refers to thermal energy transferred from a hotter to a cooler system.

The temperature of a geographic area can be affected by several factors:

- **Higher latitudes typically have colder temperatures and a wider range of temperatures.** Imaginary lines called *lines of latitude* circle the Earth east to west, indicating the distance of a location north or south of the equator.
  - **The equator is latitude zero.** The poles are at latitudes 90 degrees north and 90 degrees south.
  - **Lower latitudes are closer to the equator.** Latitudes 10 degrees north and 10 degrees south are close on either side of the equator.
  - **Higher latitudes are closer to the poles.** Latitude 80 degrees north is close to the North Pole, and latitude 80 degrees south is close to the South Pole.
- **Areas near large bodies of water have milder temperature changes.** Water heats up and cools down more slowly than air. So, water can heat up during the day, then slowly release that heat at night to nearby areas.
- **Temperatures decrease with increasing altitude.** As warm air rises, it enters areas of lower atmospheric pressure, which causes it to expand and cool.
- **Dark areas absorb radiation, while light areas reflect it.** Urban areas may be dark in color, creating *heat islands* that absorb more heat. Snow reflects radiation, helping keep the temperature at the poles cooler.

Letting in the light

Climate varies in response to the amount of energy received from the Sun. The spherical shape of the Earth affects the way light hits different latitudes, as shown in Figure 2-1:

- Near the equator, light rays strike the Earth's surface in concentrated, direct rays.
- Toward the poles, the light rays become slanted, spreading the light over a larger area.

REMEMBER

Latitudes further away from the equator receive less solar radiation. This pattern explains why temperatures are higher at the equator and decrease toward the poles.

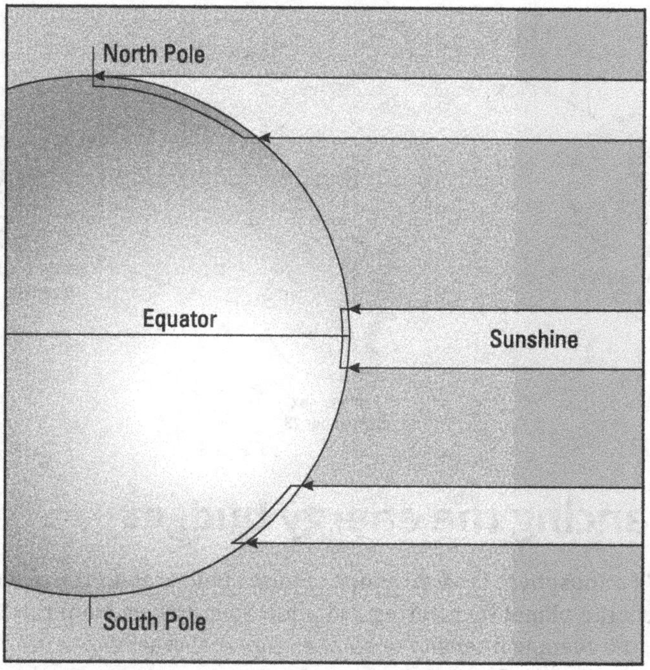

FIGURE 2-1: Angles of sunlight at different latitudes.

Solar radiation strikes the hemispheres of the Earth differently as it orbits the Sun, resulting in seasonal climate changes. Figure 2-2 shows how the tilt in Earth's axis causes the different hemispheres to be angled toward and away from the Sun at different points in its yearly orbit:

- A vertical line from the north to the south pole represents the axis around which the Earth spins. Instead of being perpendicular to the plane of Earth's orbit, the axis is tilted 23.5 degrees.
- At the June solstice, the Northern hemisphere is angled toward the Sun and the Southern hemisphere is angled away. The Northern hemisphere receives

more solar radiation and experiences the warmer temperatures of summer, while the Southern hemisphere has its winter.

» At the December solstice, the Southern hemisphere angles toward the Sun and experiences summer, while the Northern hemisphere angles away and has its winter.

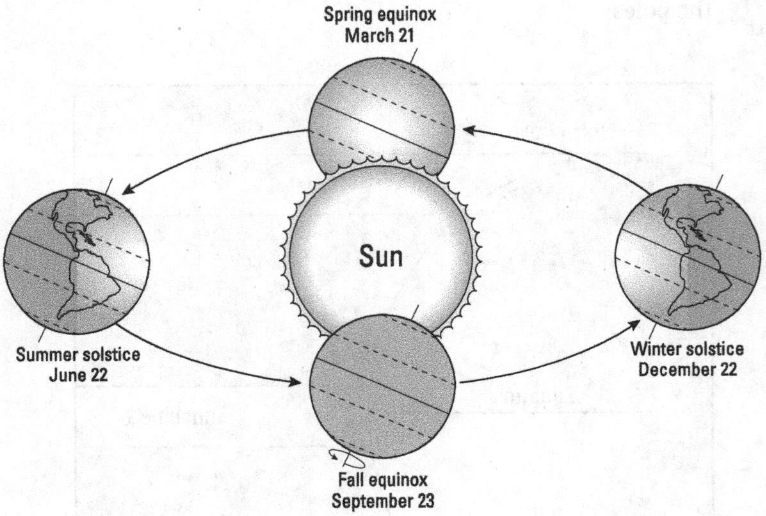

FIGURE 2-2: The Earth's tilt and orbit create seasons. Seasons labeled here refer to the Northern hemisphere.

Balancing the energy budget

Earth's atmosphere is an envelope of gases held around the planet by gravity that protects the planet by filtering radiation as it arrives from the Sun. According to NASA, the average intensity of solar energy that reaches the top of the atmosphere facing the Sun is about 1,360 watts per square meter at Earth's average distance from the Sun. Scientists call this the total *solar irradiance*.

Clouds, atmospheric particles, and the Earth's surface reflect about one-third of the total irradiance back out to space. Clouds, ozone, and water vapor absorb about one-fifth into the atmosphere. Land and water at the Earth's surface absorb the rest.

REMEMBER

The reason that Earth doesn't just continuously heat up is that the Earth's surface and the lower atmosphere constantly release energy back to space. Earth's *energy budget* is the net flow of energy into and out of Earth. For a stable climate, the energy budget of the Earth must be balanced.

Energy leaves the earth through radiation, convection, and conduction:

- ***Radiation* is the transfer of heat energy through space by electromagnetic energy.** Terrestrial radiation leaves the Earth's surface as long energy waves, transferring heat to the atmosphere or back out to space. If you've ever felt the heat from a fire, you've experienced heat transfer by radiation.
- ***Conduction* is the transfer of heat energy between two substances that are touching when the kinetic energy of molecules in one directly transfers to molecules in the other.** Molecules in solids and liquids are closer together than those in gases, so most energy transfer by conduction occurs right near Earth's surface. For example, sunlight heats the ground during the day by radiation, then the ground heats the air directly above it via conduction. If you've ever felt heat transfer from a warm mug to your hand, you've experienced conduction.
- ***Convection* is the transfer of heat energy in a fluid.** When air or water absorbs heat, the molecules expand, and the fluid rises. The warm fluid transfers heat to its surroundings, cooling and condensing so that it sinks back down. If you've ever watched a lava lamp, you've seen convection in action.

Blanketing the Earth with the greenhouse effect

As energy radiates from the Earth's surface, it warms the atmosphere from below. Some gases in the atmosphere can absorb this energy and re-radiate it back toward the Earth. Without these *greenhouse gases*, scientists estimate that Earth's climate would be 59°F (33°C) cooler than it is today. Greenhouse gases include water vapor, carbon dioxide, nitrous oxide, and methane. Increased production of greenhouse gases by humans is altering the energy budget by keeping more heat in the system. The resulting global climate change has the potential for enormous ecological consequences. (See Chapter 12 for details.)

Moving Energy via Air and Water

Scientists sometimes refer to the Earth's climate as a giant heat engine that drives the movement of the atmosphere and the oceans. In a heat engine, the transfer of heat from a hot reservoir to a cold one produces mechanical energy that may then be used to perform work. On Earth, solar radiation creates warmer regions around the equator that transfer their heat to cooler areas that are higher in latitude or altitude. This heat transfer produces work in the form of winds and ocean currents.

REMEMBER

Large horizontal circulation systems, like winds and ocean currents, move energy from the equator to the poles of the Earth.

Forming circulation cells

Solar radiation arrives at higher latitudes on the Earth at an oblique angle, as shown in Figure 2-1. This radiation has to penetrate more atmosphere than radiation arriving at the equator, which means there's more opportunity for scattering and reflection of the energy. Thus, Earth's atmosphere contributes to the energy imbalance between the equator and the poles.

Figure 2-3 shows how the higher energy at the equator is dispersed toward the poles by convection in the atmosphere:

- » The Sun delivers the greatest intensity of energy to the Earth at the equator. The Earth's surface warms, then radiates energy back to the air.
- » Warm air is lighter and less dense than cool air, so it rises in a process called *uplift*.
- » When the warm air reaches the boundary between the lower atmosphere, or troposphere, and the next atmospheric layer, the stratosphere, it flows toward the poles.
- » As the air moves, it transfers heat to its surroundings, becoming cooler and more dense. Eventually, it sinks back toward the surface in a process called *subsidence*.

REMEMBER

The uplift, flow, and subsidence of air near the equator create a three-dimensional pattern of atmospheric circulation called a *Hadley cell*.

Similar circulation cells occur within the latitudes between the equator and the poles, as shown in Figure 2-3. Together, these circulation cells establish the major climatic zones on Earth:

- » The *tropical zone* is between the equator and 30 degrees north and south.
- » The *temperate zones* are between 30 and 60 degrees north and south.
- » The *polar zones* are above 60 degrees north and south.

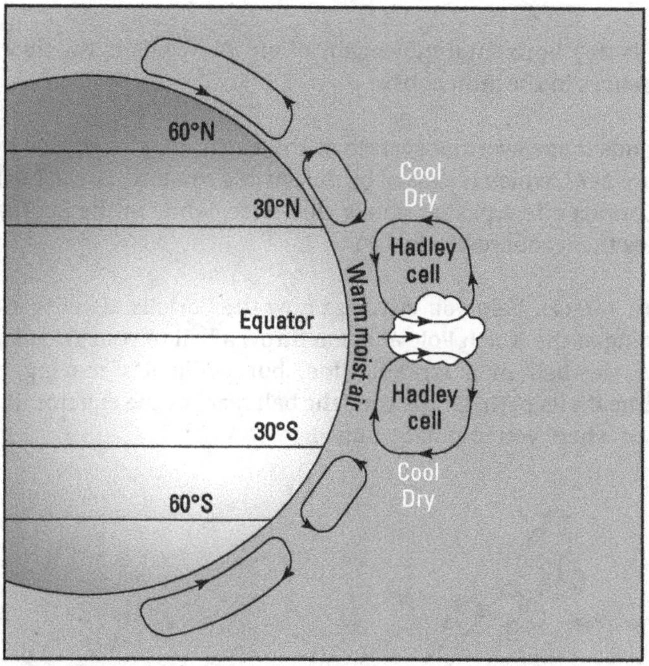

FIGURE 2-3: Hadley cells.

In addition to moving energy through the atmosphere, circulation cells also affect precipitation:

- As warm air rises, it creates *low-pressure* zones. As the rising air cools, the water it carries condenses, forming clouds that may release as rain. Low-pressure areas, like those around the equator, typically get lots of rain in all seasons.
- As cool air sinks, it creates *high-pressure* zones. Sinking air compresses, making the air warm. Water in warm air doesn't condense into liquid water. High-pressure areas, like those at the poles and at 30 degrees N and S, don't usually get a lot of precipitation.

Major air currents

The difference in air pressure in the circulation cells creates a *pressure gradient* that causes air to move from the high-pressure to the low-pressure areas. Air moves in consistent patterns across the surface of the Earth, known as prevailing winds.

REMEMBER

Wind is any horizontal movement of air. As winds blow, they equalize pressure differences in the atmosphere.

As winds move over the surface of the Earth, they're shifted by the *Coriolis effect* (Figure 2-4), which is caused by the Earth's rotation. In the Northern hemisphere, the Coriolis effect pushes winds clockwise, while in the Southern hemisphere, it pushes them counterclockwise.

TIP

Figure 2-4 can help you visualize how the Coriolis effect works. Imagine you're standing at the North Pole and you throw a ball to your friend at the equator. You throw the ball in a straight line, but while it's moving, the Earth rotates underneath its path. By the time the ball reaches the equator, it lands in a location west of where you originally aimed.

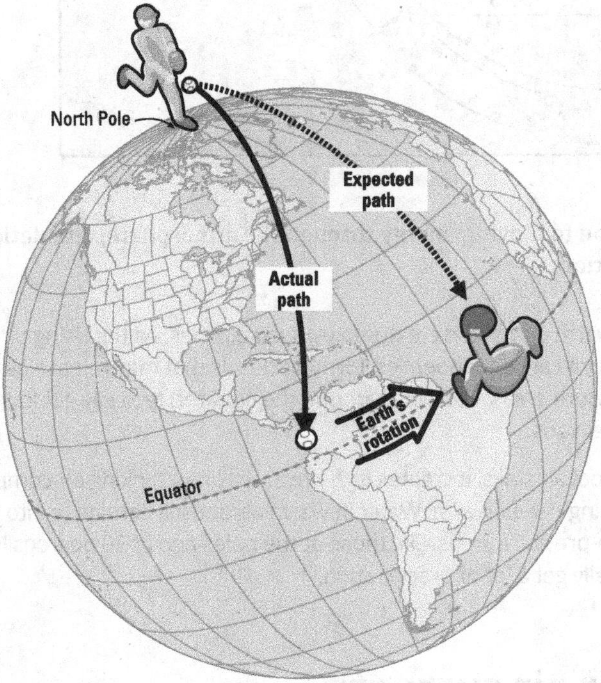

FIGURE 2-4: The Coriolis effect.

Figure 2-5 shows the prevailing winds and how they're shifted by the Coriolis effect. Winds are named for their direction of origin:

- The *easterlies* blow east to west from the subtropical high-pressure zones toward the equator. These winds got the name trade winds because of their importance to trade in the era of sailing ships. (To learn more about how the trade winds influence climate, see the sidebar "El Niño y La Niña.")

- The *westerlies* blow west to east from subtropical high-pressure zones toward the poles.
- *Jet streams* are narrow bands of strong wind high in the atmosphere that blow west to east. They result from multiple factors, including the Earth's rotation and the meeting of cold and warm air masses. Jet streams can move storms very quickly over the Earth's surface.

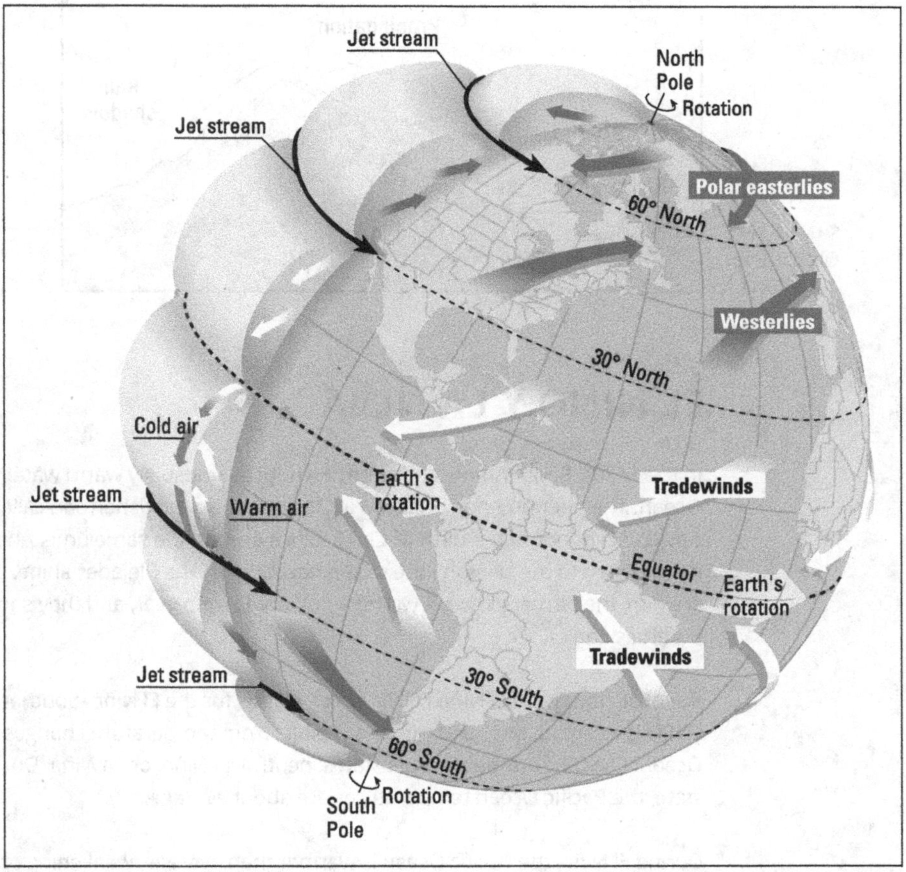

FIGURE 2-5: The general circulation of the atmosphere.

Other factors influence the movement of winds, including temperature changes due to the seasons and local geographic features like mountains. The movement of air over mountains causes *orographic effects*, as shown in Figure 2-6:

1. Warm, moist air begins to flow over a mountain range.
2. As the air rises over the mountain, it cools, causing water to condense and fall as precipitation.
3. Dry air flows down the other side of the mountain, warming as it descends. The warm, dry air causes desert-like conditions called a *rain shadow* effect.

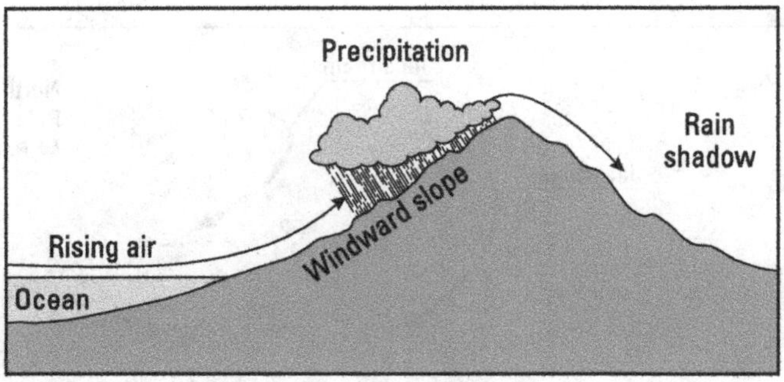

FIGURE 2-6: Orographic effects.

EL NIÑO Y LA NIÑA

In the 1600s, South American fishermen noticed unusually warm water in the Pacific Ocean. The phenomenon recurred in December, so the fisherman called it *El Niño de Navidad* in honor of the birth of Christ. Since around the same time, Andean farmers have looked to the skies in June to see how brightly the Pleiades shine. If the Pleiades are dim, the farmers know it will be a hot, dry El Niño year, and they should plant their potatoes earlier.

Scientists today call El Niño ENSO, which stands for the El Niño–Southern Oscillation. ENSO is a climate phenomenon that results from temperature changes in the Pacific Ocean. ENSO has three possible states: neutral, El Niño, or La Niña. During the neutral state, the Pacific Ocean temperatures are about average.

During El Niño, the Pacific Ocean is warmer than average, weakening or even reversing the trade winds. Warm water is pushed east, toward the west coast of the Americas. This shift causes areas in the northern U.S. and Canada to be dryer and warmer than usual. In contrast, areas in the U.S. Gulf Coast and Southeast are wetter than usual and have increased flooding. Increased lightning and tornado activity also occur in El Niño years.

The opposite of El Niño is La Niña. During La Niña events, the Pacific Ocean is cooler than average, increasing the force of the trade winds. The winds push warm water toward Asia, increasing upwelling off the west coast of the Americas. The cold waters in the Pacific push the jet stream northward, causing drought in the southern U.S. and heavy rains and flooding in the Pacific Northwest and Canada. The Southern United States may be warmer than usual, while the Northern U.S. is cooler. La Niña can also cause a more severe hurricane season.

Major water currents

As winds move across the surface of oceans, they create a frictional drag that moves the water. As a result, the pattern of ocean surface currents is similar to that of the prevailing winds. Current patterns differ, however, because ocean currents bump up against land masses that winds can blow right over. Also, the Coriolis effect causes some ocean currents to break up into circular vortices called *eddies*.

REMEMBER

An *ocean current* is the horizontal movement of water.

The winds that form from high-pressure areas at latitudes 30 degrees north and south help form circular ocean circulation patterns called *gyres*, shown in Figure 2-7. Gyres form from a combination of wind hitting a land mass and the Coriolis effect, resulting in currents that rotate in a clockwise direction in the Northern hemisphere and a counterclockwise direction in the Southern hemisphere.

Ocean gyres flow around large areas of calm, stable water. The circular motion of the gyre draws garbage into this stable center, where it can become trapped for years. Much of the garbage in these patches are plastics that doesn't decompose and is harmful to marine life.

Like the movement of air, the movement of water around the Earth also has a vertical dimension, which is driven by density and pressure:

- Water closer to the equator absorbs more energy and is warmer. Warm water expands and is less dense than cold water.
- The density of water is also affected by the amount of salt. Density increases with increasing salt.

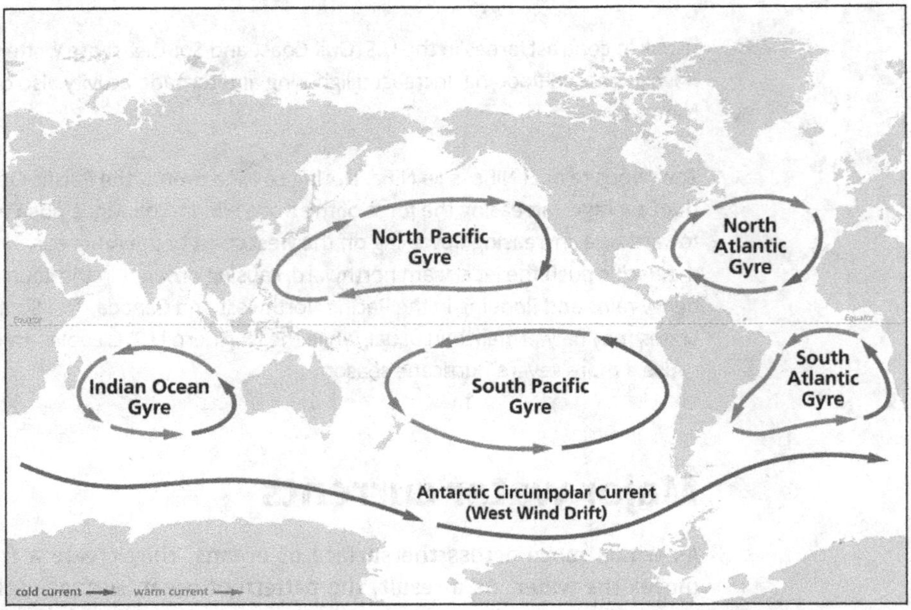

FIGURE 2-7: Five of the world's ocean gyres.

Peter Hermes Furian Adobe Stock Photos

Surface currents generally move cold water toward the equator. Surface and deep ocean waters typically don't mix because surface waters are warmer, less dense, and have a lower salt concentration than deep water. As surface water reaches the poles, however, it cools and becomes denser. The formation of ice draws water out of solution, leaving denser, saltier water behind. This cold, dense water sinks to deeper layers and begins to flow back toward the equator. The sinking of surface water brings oxygen into the depths where organisms can use it. Because the flow of water from the equator to the poles and back again relies on both temperature and salt concentration, it's called *thermohaline circulation* (*thermo* = heat; *halo* = salt).

REMEMBER

The thermohaline circulation of ocean water, shown in Figure 2-8, creates an interconnected system of surface and deep ocean waters that transfers energy from the equator to the polar regions. This system is also sometimes called the great ocean conveyor belt.

In areas where prevailing winds blow surface water away from the land, deep water may rise to replace it in a process called *upwelling*. Upwelling is important because it brings nutrients from deep water up to the surface, making them available to organisms that live there.

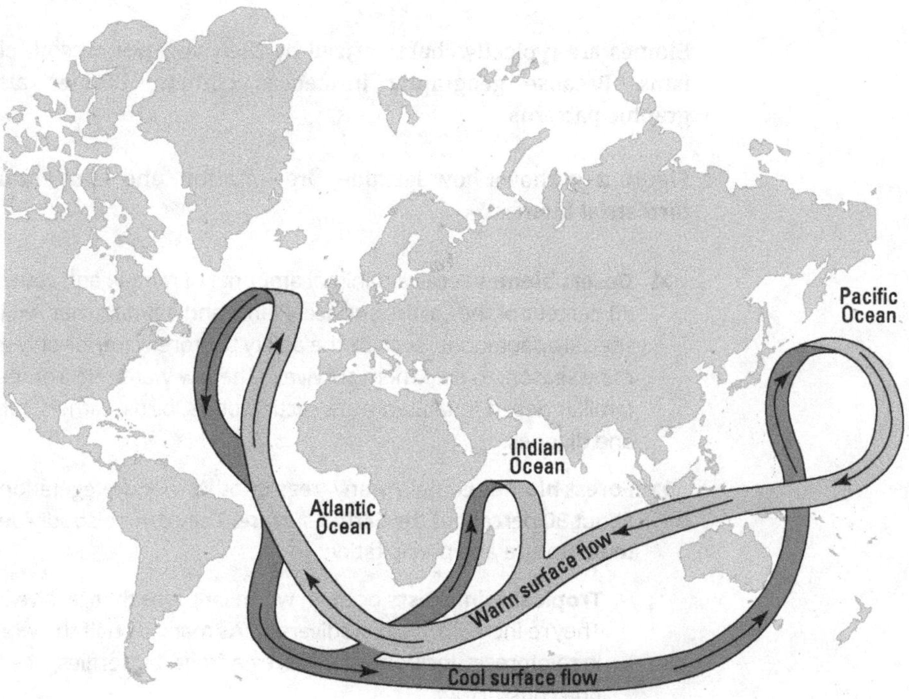

FIGURE 2-8: Thermohaline circulation of ocean water.

Examining the Major Biomes

A German climatologist named Wladimir Köppen divided the world's climates into major categories based upon their general temperature profiles and latitude. His original five categories were tropical, dry, temperate, continental, and polar climates. Working with another climatologist, named Rudolph Geiger, Köppen further divided his climate scheme based on precipitation, seasonal differences, and altitude, resulting in 31 categories.

The striking difference between the categories is the type of vegetation that grows in each. Two areas that are completely separated in space but that belong to the same category have very similar types of species. For example, species that grow in the tropical rainforests of South America, Africa, and Asia are barely related to each other, but they have very similar structures due to convergent evolution (for details on convergent evolution, see Chapter 3). The similarity of these unrelated species demonstrates the impact of climate on communities and evolution.

REMEMBER

A *biome* is a community that's adapted to a specific climate.

Biomes are typically characterized by their vegetation, soil, climate, and organisms. Because geography influences climate, biomes also illustrate geographic patterns.

Figure 2-9 shows how latitude, precipitation, and temperature create different terrestrial biomes:

- **Desert biomes** receive minimal amounts of rainfall and cover approximately 20 percent of the Earth's surface. Plants and animals that live in deserts have special adaptations, such as the ability to store water or only grow during the rainy season, to help them survive in the low-water environment. Some familiar desert inhabitants are cacti, reptiles, birds, camels, rabbits, and dingoes.
- **Forest biomes** contain many trees or other woody vegetation and cover about 30 percent of the Earth's surface. They can be subdivided based on temperature and precipitation:
  - **Tropical rainforests** occur in warm latitudes that receive lots of rainfall. They're incredibly rich biodiversity. As many as half the world's animals live in rainforests, including gorillas, tree frogs, butterflies, tigers, parrots, and boa constrictors.
  - **Temperate rainforests** are found in temperate regions that receive lots of rainfall. They may be dominated by broad-leafed or coniferous trees.
  - **Taiga,** or boreal forests, occur in cold regions and are the largest terrestrial biome. They're coniferous forests made up mostly of spruce, pine, and larch trees.
- **Grassland biomes** are dominated by grasses because the low precipitation supports grasses but not trees. Grasslands cover about 30 percent of the Earth's surface.
  - **Savannahs** occur in areas that have warm climates with a rainy season and a dry season. Savannas in Africa are home to zebras, giraffes, elephants, lions, and hyenas. Savannahs can also be found in South America, Asia, and Australia.
  - **Temperate grasslands** have rich soil and abundant grasses. They are home to species like prairie dogs and buffalo.
- **Tundra biomes** are very cold and have very little liquid water. Tundras cover about 15 percent of the planet's surface and are found at the poles of the Earth as well as at high elevations. Arctic tundras are home to arctic foxes, caribou, and polar bears, whereas mountain tundras are home to mountain goats, elk, and birds. In both types of tundra, nutrients are typically scarce, and the growing seasons are quite short.

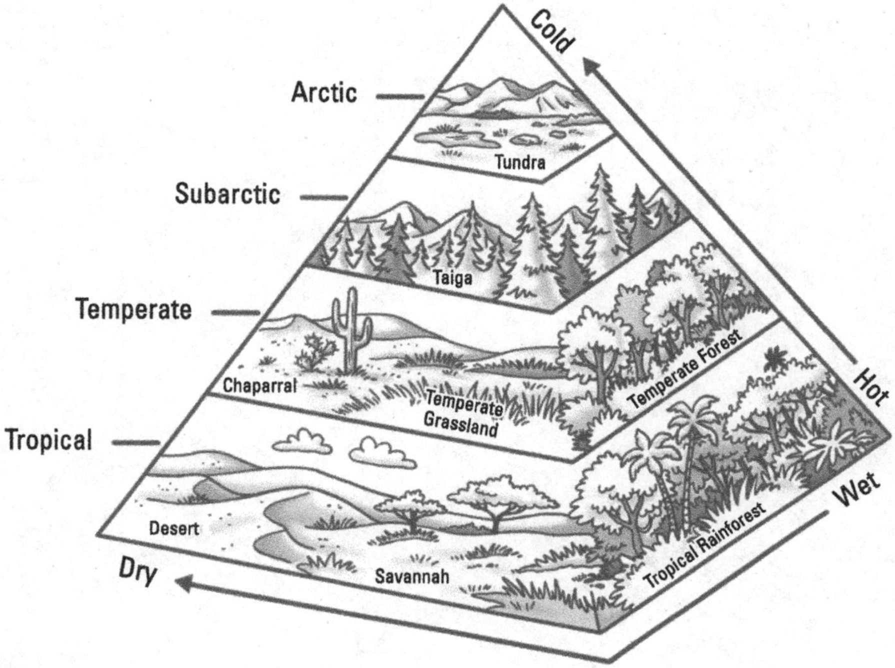

FIGURE 2-9: Some terrestrial biomes.

- **Aquatic biomes** include both freshwater and marine biomes:
  - **Freshwater biomes** include ponds, rivers, streams, lakes, and wetlands. Only about 3 percent of the Earth's surface is made up of freshwater, but freshwater biomes are home to many different species, including plants, algae, fish, and insects. Wetlands, which are semi-aquatic biomes where water covers the soil, have the greatest amount of diversity of any of the biomes.
  - **Marine biomes** contain saltwater and include the oceans, coral reefs, and estuaries. They cover 75 percent of the Earth's surface and are very important to the planet's oxygen and food supply — more than half the photosynthesis that occurs on Earth occurs in the ocean. Marine biomes are home to many marine creatures such as algae, fish, octopuses, dolphins, and whales.

IN THIS CHAPTER

» Considering the pros and cons of mutations

» Monitoring genetic changes in populations

» Examining change over long timescales

Chapter 3
Evolving in a Changing World

Living things are amazingly diverse yet also have interesting similarities. Dolphins and sharks share streamlined bodies and fins, yet one is a mammal, and one is a fish. On the surface, a tropical rainforest in South America looks very similar to one in Africa: both have tall broad-leafed trees that host climbing vines and clinging epiphytes. A closer look reveals that the similar-looking plants in each forest are actually different species. Similar environmental conditions in both locations sculpted the species over time so that they look and behave in similar ways.

REMEMBER

Convergent evolution occurs when unrelated organisms develop similar features or behaviors, often as solutions to the same environmental challenges.

In this chapter, you'll explore the intersection between ecology and evolution, from the processes that generate diversity to those that shape groups of organisms.

Endless Variety

An engine that generates diversity quietly hums in the background of all living things. Every time cells multiply, whether for reproduction, growth, or repair, tiny changes creep into the genetic code that determines their traits.

REMEMBER

A *mutation* is a change in the genetic code.

The code in DNA is spelled out in chemical subunits, called *nucleotides*, that scientists represent with the letters A, C, G, and T. Every time a cell divides, an enzyme called *DNA polymerase* copies the genetic code so that the new cells get a copy. DNA polymerase is very good at its job, but it isn't perfect. Every time it copies DNA, it makes some mistakes. For example, scientists estimate that every human baby has 60 to 70 new mutations that weren't present in their parents.

Spontaneous mutations occur all the time because of the built-in error rate of enzymes that copy genetic material. Exposure to certain environmental agents can increase the rate of mutation, causing *induced mutations*.

REMEMBER

A *mutagen* is anything that increases the mutation rate. Examples of mutagens include ultraviolet radiation from the sun and polycyclic aromatic hydrocarbons (PAH) produced during the burning of fossil fuels.

Mutations have a range of effects on individual organisms, from no effect at all to severe consequences. The impact depends upon many factors, including whether the organism is single-celled or multicellular or whether the mutation affects an important protein. Although some individuals may be negatively affected by a mutation, most survive and carry small changes in their genetic code.

REMEMBER

Genetic variation is the genetic difference between individuals of the same species.

Everyday Evolution

Evolution leads to changes in the traits of populations over time. An example that you've probably heard about is the evolution of antibiotic-resistant bacteria. Bacteria reproduce very rapidly, so their generations fly by, making it easier to see the impact of evolution during a human lifespan. Since people started using antibiotics in the 1940s to treat infections, some populations of bacteria have changed so that an antibiotic that once killed them no longer works. This is an example of evolution in action because the original trait that was most common, susceptibility to an antibiotic, is less common. A very rare trait, being resistant to the antibiotic, has increased in the population.

REMEMBER

Evolution is the change in the frequency of traits of populations over generations.

Reviewing key concepts

Scientists can follow evolution in action through *population genetics*, which looks at the underlying changes in genetic structures that determine traits in populations. When thinking about population genetics, it's helpful to know a few fundamentals about genes and how they work:

- **The entire genetic code of an organism is its *genome*.**
- **Genomes contain blueprints called *genes* that determine the traits of organisms.** Most genes are the instructions for building a specific protein. The way proteins do their jobs determines traits. A mouse might have genes that affect the color of its fur, for example.
- **Different variations of a gene are called *alleles*.** In a population of mice, some individuals might have alleles of the fur color gene that specify white fur, while others have alleles that cause brown fur to grow.
  - Some alleles can hide the presence of other alleles. For example, the allele for brown fur might contain the code for a protein necessary to make brown pigment. The allele for white fur might be dysfunctional, so that no pigment is made. If a mouse has at least one copy of the functional allele, it will be brown even if its other allele is dysfunctional because some brown pigment is enough to make a brown mouse. Alleles that can hide other alleles are *dominant*. Alleles that can be hidden are *recessive*.
  - A common shorthand is to use a letter of the alphabet to represent a gene, and to use lower- and upper-case letters to represent alleles. If I choose the letter R to represent the mouse fur color gene, then *R* would represent the dominant allele for brown pigment, and *r* would represent the nonfunctional allele.
- **The specific combination of alleles that an organism has is its *genotype*.**
  - Many organisms, including mice and humans, have two copies of every gene, except for genes on the sex chromosomes in males.
  - Using the letter R to represent alleles, different mice could have the genotypes *RR*, *Rr*, or *rr*.
- **The combination of traits that an organism displays is its *phenotype*.** In this example, a mouse could have the phenotype of white fur or brown fur. The mice with brown fur could have the RR or Rr genotype, while white mice will always have the rr genotype.

Tracking gene flow

The *gene pool* is all the possible alleles of the various genes that can be found among the individuals in a population. As individuals reproduce sexually, new combinations of alleles may occur in offspring, even if the numbers and types of alleles in the gene pool remain the same. Population geneticists track genetic changes between generations in two ways, which are illustrated in Figure 3-1:

- **»** ***Genotype frequencies*** **represent how common each genotype is in a population.** Imagine that the circles in Figure 3-1 represent brown and white mice (each circle represents 10 mice). Some brown mice have the genotype *RR*, while others have the genotype *Rr*. White mice have the genotype *rr*.
  - To calculate the observed genotype frequencies shown for the present generation on the left in the figure, you count up the number of individuals of each type, then divide by the total number in the population. You can multiply by 100 to convert the proportion to a percentage. The genotype frequency for *RR* is $60/130 \times 100 = 46$ percent. For *Rr*, $50/130 \times 100 = 39$ percent, and for *rr* $20/130 \times 100 = 15$ percent.
  - When parent mice reproduce to make the next generation, they copy their alleles and give one copy of each gene to their sperm or egg. All of the possible alleles from the parents are shown in the gene pool in the middle of the figure.
  - When sperm and egg combine, parental alleles come together in new combinations. These new combinations are represented in the changed genotype frequencies of the next generation on the right in the figure: 38 percent *RR*, *54* percent *Rr*, and 8 percent *rr*.
- **»** ***Allele frequencies*** **represent how common a particular allele is in a population.** To calculate observed allele frequency, you divide the number of copies of a particular allele in the population by the total number of all the alleles of that gene in the population. If you know the genotypes of a population, you can count the number of each allele and then divide by the total. A mouse that is *RR* has two copies of the *R* allele; a mouse that is *Rr* has one copy of the *R* allele. Likewise, a mouse that is *rr* has two copies of the *r* allele, and a mouse that is *Rr* has one copy of the *r* allele.
  - In the present generation on the left in Figure 3-1, the six dark circles represent 60 mice that have the genotype *RR*, for a total of 120 *R* alleles. Five half-shaded circles represent 50 *Rr* mice, so that adds another 50 *R* alleles for a total of 170 *R* alleles in the population. Thirteen circles indicate a population of 130 individuals, which means 260 total alleles. To determine the frequency of *R*, you divide 170 by 260, which equals 0.65.

- To calculate the observed frequency of the *r* allele, first look at the two circles that represent 20 light-colored mice. These mice have two *r* alleles each, for a total of 40 *r* alleles. The *Rr* mice contribute another 50 *r* alleles for a total of 90. Divide 90 by 260, and you get 0.35 as the observed frequency of the *r* allele.

REMEMBER

Population geneticists represent the frequency of the dominant allele with the letter *p* and the frequency of the recessive allele with the letter *q*. For a gene with two alleles, $p + q = 1$. This equation allows you to calculate the frequency of one allele if you know the other. For example, if you know that the frequency of $q = 0.25$, then the frequency of *p* must be 0.75 because $p = 1 - q$. Or, for the example of mice in Figure 3-1, you can see that $0.65 + 0.35 = 1$.

REMEMBER

Gene flow is the movement of individuals and their genetic material from one population to another.

If you compare the present generation with the next generation shown in Figure 3-1, you can see that the present generation has 20 white mice and the next generation only has ten. If you were observing this population of mice and only looking at phenotypes, you might think that the genetic information that determines white fur color decreased in the population from one generation to the next. Because the figure shows you the genotypes and alleles, though, you can see that this isn't true. The allele that determines white fur color is still present at the same frequency in the next generation; it's just hidden in the heterozygotes.

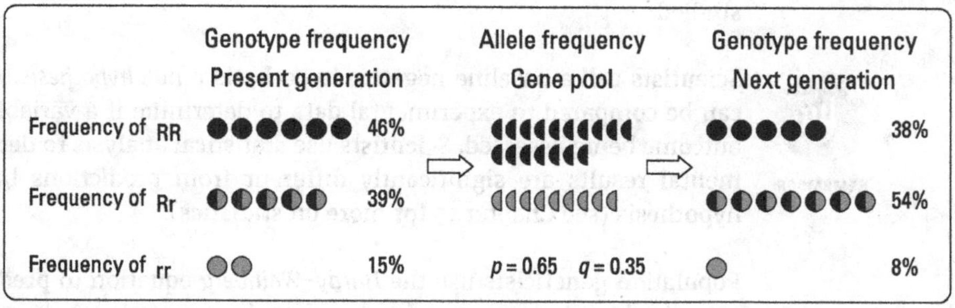

FIGURE 3-1: Allele and Genotype Frequencies.

Testing for equilibrium

Population geneticists use allele and genotype frequencies to study evolution on the genetic level. Two scientists, named G.H. Hardy and W. Weinberg, created a mathematical model that uses allele frequencies to predict the genotype frequencies you would expect to see in a population that is in equilibrium, meaning that it is not evolving.

REMEMBER

For a population to reach the *Hardy-Weinberg equilibrium*, several things must be true:

- The population must be very large.
- New alleles aren't entering the population through mutation or migration.
- Mating is random.
- Alleles have the same survival advantage.

If all of these conditions are met, a population can reach the Hardy-Weinberg equilibrium in one generation. When a population is in equilibrium, the genotype frequencies can be modeled with the *Hardy-Weinberg equation*:

$$p^2 + 2pq + q^2 = 1$$

In this equation:

- Homozygous dominant individuals, like mice that are *RR*, are represented by p^2.
- Heterozygous individuals, like mice that are *Rr*, are represented by $2pq$.
- Homozygous recessive individuals, like mice that are *rr*, are represented by q^2.

The Hardy-Weinberg equation is useful because it sets a baseline expectation for genotype frequencies if evolution isn't occurring to influence the gene being studied.

REMEMBER

Scientists call a baseline negative hypothesis a *null hypothesis*. A null hypothesis can be compared to experimental data to determine if a variable is affecting the outcome being assessed. Scientists use statistical analysis to determine if experimental results are significantly different from predictions based on the null hypothesis (see Chapter 15 for more on statistics).

Population geneticists use the *Hardy-Weinberg* equation to predict genotype frequencies based on the assumption that the population is in equilibrium. Then, they can compare their predictions with observations of real populations and see if the numbers match. If they don't, then something is causing the population to evolve. The difference between the observed and predicted genotype frequencies can help population geneticists make additional hypotheses about how evolution is affecting the population and indicate directions for additional experiments.

As an example of how population geneticists use the Hardy-Weinberg equation, you can use the allele frequencies for the present generation on the left in Figure 3-1 to make predictions for what the genotype frequencies would be if that population is in equilibrium:

1. You already used the information for the population to calculate the observed allele frequencies for *p* and *q*: $p = 0.65$ and $q = 0.35$.
2. From this, you can calculate the expected genotype frequencies if this population is in equilibrium as $p^2 = 0.4225$, $q^2 = 0.1225$, and $2pq = 0.4555$. Notice that these three numbers do add up to one. You can multiply these numbers by 100 to create percentages.
3. When you compare these predictions to the observed genotype frequencies on the left in Figure 3-1, you can see that they don't match.
   a. The prediction for p^2, or *RR*, is 42.25 percent, but the observed is 46 percent.
   b. The prediction for $2pq$, or *Rr*, is 45.55 percent, but the observed is 39 percent.
   c. The prediction for q^2, or *rr*, is 12.25 percent, but the observed is 15 percent.
4. The comparison between Hardy-Weinberg predictions and observations shows that this population isn't in equilibrium, suggesting that evolution may be acting on the population. In particular, the number of heterozygotes is lower than predicted, which might indicate some disadvantage to being heterozygous for this gene.

TECHNICAL STUFF

Population geneticists normally use a statistical test called the *chi-square goodness-of-fit test* to assess the significance of differences between observed and expected genotype frequencies based on the Hardy-Weinberg equilibrium. (For more on statistics, check out Chapter 15.)

The Hardy-Weinberg equilibrium can only be achieved in populations in which alleles aren't changing. Depending upon the gene being studied, this condition often isn't met in real populations. One way that allele frequencies in a population can change is through migration. If individuals move into or out of a population, they take their alleles with them. Allele frequencies in a population can also change through mutation, natural selection, genetic drift, and non-random mating. Natural selection and genetic drift are presented later in this chapter.

REMEMBER

Population geneticists also define evolution as a change in allele frequencies over generations.

Survival of the Fittest: Natural Selection

Most people's ideas about evolution are probably based on the work of Charles Darwin. Darwin served as the ship's naturalist on the HMS *Beagle* from 1831 to 1836 as it sailed along the coast of South America and then around the world. As a naturalist, Darwin spent a great deal of time going ashore to observe the natural world. He wrote detailed notes, made sketches, and collected specimens of plants, animals, and fossils along the way.

Darwin saw many wondrous things, from flightless birds in Patagonia to swimming iguanas in the Galapagos Islands to fossils of extinct giant ground sloths in Argentina. He noticed the similarities and differences between related organisms and thought about the fact that some things, like the giant sloth, existed in the past but not in his time. He started wondering about how organisms become different from each other and how it might be that life changes over time.

The birds in the Galapagos had a big impact on his thinking. On the islands, some birds ate seeds, others ate insects, and some even ate cactuses. The beak of each bird seemed exactly suited to its food source. Darwin recognized some as finches, but others looked so different from finches that he thought they were wrens or orioles. When he got back to England and showed his specimens to a bird specialist, he was surprised to learn they were all finches. He speculated that the diverse finches could have descended from some ancestral finches that came to the Galapagos from the mainland. If finches arrived on a new island that had no other birds, he wondered if they could change over time to take advantage of all of the available resources.

To test his ideas and see if it was possible, Darwin started breeding pigeons. Pigeon fanciers breed pigeons for their speed or for their visual appeal, choosing breeding pairs for their color, fan-shaped tails, neck frills, and curly feathers.

REMEMBER

Artificial selection is when humans use selective breeding to obtain desired traits in other organisms.

Darwin compared the results of artificial selection with what he saw in nature, ultimately arriving at the idea that new species could arise from common ancestors by descent with modification, in other words, by changing through the generations. He proposed a mechanism that he called *natural selection* to explain how this change could happen.

REMEMBER

Natural selection occurs when organisms that are better suited to their environment are more likely to survive and produce offspring. This idea is commonly referred to as *survival of the fittest*.

WARNING

In ecology, fitness doesn't necessarily mean the strongest, fastest, or smartest. Fitness, and survival of the fittest, refers to an organism that has a trait that makes it reproductively successful. It could be any trait that benefits them in their particular environment.

Establishing Darwin's principles

Darwin proposed that, for natural selection to occur in a population, several things must be true:

- **Variation:** Individuals in the population must have different characteristics. Mutation creates new alleles, and sexual reproduction leads to new combinations of alleles in offspring through meiosis and random fertilization.
- **Competition:** Individuals in the population must produce more offspring than can survive, resulting in competition for resources. People can make conscious choices about how many offspring they have. Most other organisms, however, produce as many offspring as they can.
- **Inheritance:** Offspring must inherit some characteristics from their parents. Offspring inherit traits from their parents when they receive copies of their parents' genes.
- **Advantage:** Organisms with the best-suited characteristics for their environment are more likely to survive and reproduce. This is the heart of natural selection. If there's competition for survival and not all the organisms are the same, then the ones with the advantageous traits are more likely to survive. If these traits can be inherited, then the next generation will show more of these advantageous traits.

When these four conditions exist, populations may change in ways that make them distinctly different from their ancestors.

REMEMBER

Speciation is the formation of new and distinct species from an ancestral organism.

WARNING

Populations evolve, not individuals. Individuals can't change their genetic traits in response to environmental stress: they live or die, reproduce or don't, based on their traits. The survivors write the story of the next generation because it's their traits that get passed on. When this happens, a group of organisms begins to show change in response to the selection pressures of the environment.

REMEMBER

A *selection pressure* is an environmental factor that influences whether individuals survive and reproduce.

An example of evolution by natural selection is shown in Figure 3-2. In this example, a visual predator like a hawk scans the landscape for prey. Brown and white mice move across a dark-colored landscape. White mice are easier to spot against a dark background, so white mice are more likely to be caught as prey. Brown mice are more likely to survive and reproduce. Over many generations, brown mice have become more common and white mice have become rarer. The frequency of the alleles for white fur color decreases as the population evolves.

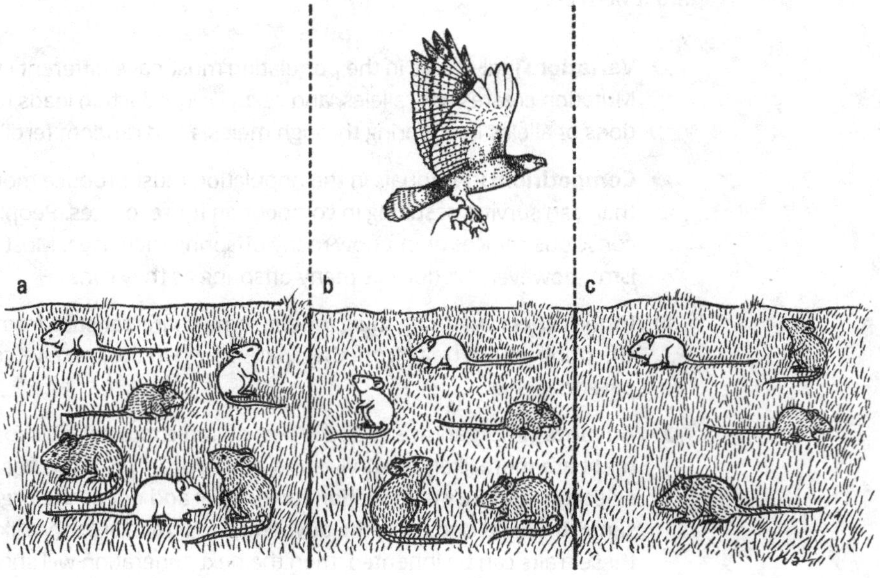

FIGURE 3-2: Evolution by natural selection.

Pushing populations in different directions

Natural selection can drive the evolution of a population, but it doesn't always push the population to change in the same way. Selection pressures change, and evolution shifts with it.

One example of this comes from peppered moths in London (see Figure 3-3). Peppered moths got their name from the tiny black speckles scattered across their mostly white wings, but some moths in the population have a rare mutation that makes them mostly black. As environmental conditions changed before and after the Industrial Revolution, the evolution of the peppered moth shifted with it:

- Before the Industrial Revolution, the peppered moth population in London consisted mostly of white moths with black speckles across their wings because this coloration gave great camouflage against lichen-covered tree trunks, helping the moths hide from predatory birds.

- During the Industrial Revolution, air pollution from the burning of fossil fuels killed the lichens, and soot from coal fires darkened tree trunks and buildings. The black peppered moths now had the advantage of better camouflage, and the population evolved so that black-colored moths became the most common.
- After the government passed laws to reduce air pollution in the middle of the 20th century, the lichens grew back again, and sooty surfaces were cleaned. Natural selection now favored the white moths, and the population evolved again.

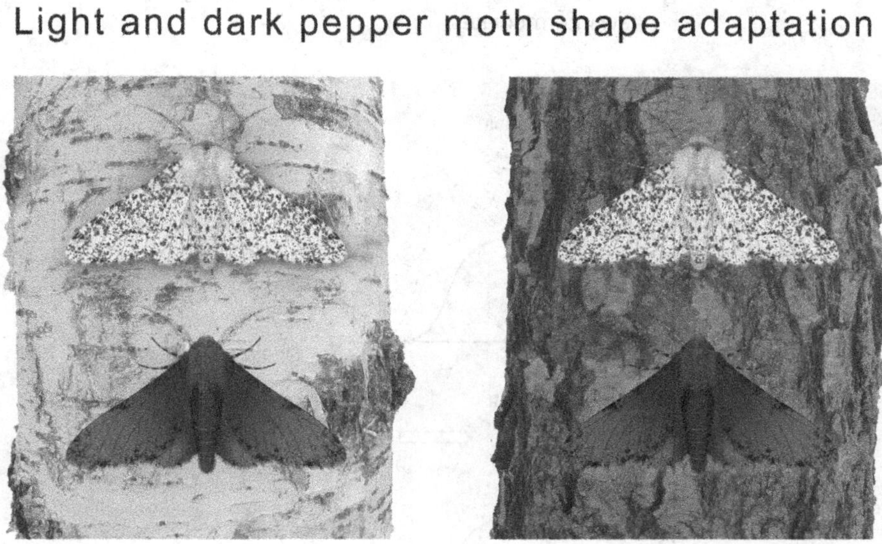

FIGURE 3-3: Peppered moths before and after the Industrial Revolution.

Aldona/Adobe Stock Photos

Figure 3-4 illustrates how natural selection shapes populations, depending upon the selection pressure:

- **Directional selection favors individuals with phenotypes at one end of the range.** The example of a visual predator that favors one phenotype of prey over another, as shown in Figure 3-2, is an example of directional selection. The dark-colored mice in Figure 3-2 are more likely to survive, leading to a gradual shift in the population toward dark fur. A real-life example of this occurs with populations of rock pocket mice in the American Southwest. Populations of mice that live on light-colored sand have light-colored fur, whereas populations of mice that live on dark lava rocks have dark-colored fur.

» **Stabilizing selection favors individuals with phenotypes in the middle of the range.** The number of eggs laid by American robins shows evidence of stabilizing selection. Robins that lay too few eggs may not have offspring that survive to adulthood. Robins that lay too many may not be able to feed all their chicks adequately. Stabilizing selection has favored robins that lay three to five eggs, with four being the most common number.

» **Disruptive selection favors individuals with phenotypes at both extremes of the range.** African seedcracker birds rely on two different plants in their environment, one that makes hard seeds and one that makes soft seeds. Seedcrackers with large beaks have an advantage when feeding on the hard seeds, while those with small beaks have an advantage with the soft seeds. As a result of disruptive selection, beak sizes in seedcrackers show a bimodal distribution, with most birds having large or small beaks, with very few intermediates.

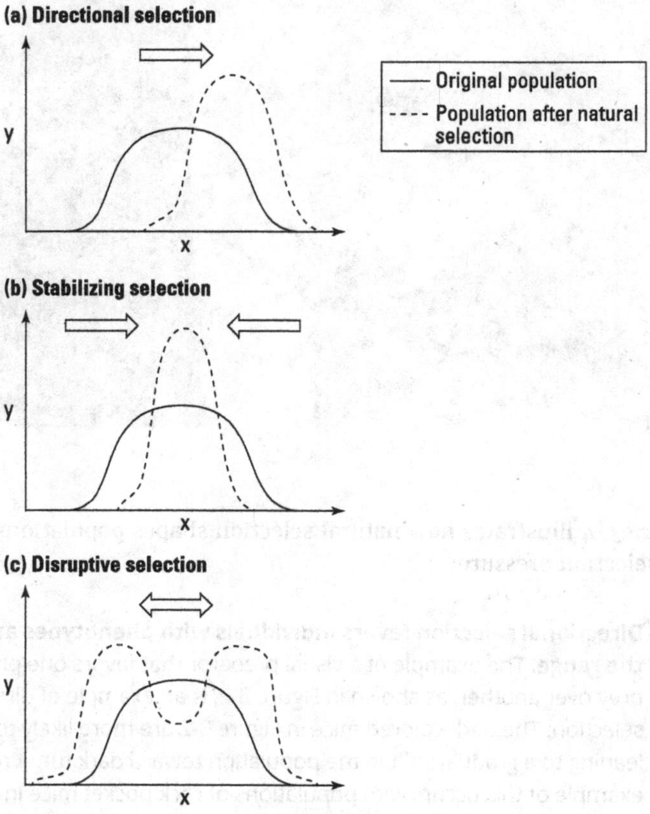

FIGURE 3-4: Types of natural selection.

Changing with Random Events: Genetic Drift

Natural selection isn't the only factor that shapes the evolution of populations. Sometimes evolution is directional, but sometimes it's random.

REMEMBER

Genetic drift is the change in allele frequencies in a population due to random chance.

A tragic example of genetic drift happened in Norway in 2016 when a lightning strike killed over 300 reindeer at once. When these reindeer died, their alleles were instantly removed from the larger population of reindeer in the area. This wasn't a case of survival of the fittest; it was just bad luck.

REMEMBER

Because genetic drift randomly removes alleles from populations, it has a larger impact on small populations than large ones.

Two types of genetic drift are shown in Figure 3-5:

- ***Bottleneck effects* occur when natural disasters or human effects drastically reduce a population's size.** Genetic diversity shrinks along with the population, shifting the allele frequencies in subsequent generations. In the late 1800s, humans hunted Northern elephant seals almost to extinction, until there were fewer than 100 individuals around 1910. After hunting stopped, the population grew back to its earlier size. Today, about 150,000 individuals exist, mostly along the coast of California. Even though the population survived the bottleneck, they're less genetically diverse than Southern elephant seals that didn't experience the same levels of hunting.
- ***Founder effects* occur when a new population begins from just a few individuals from the original population.** The new population is likely to have lower genetic diversity than the original population and to have different allele frequencies. The Mexican gray wolf population in the Southeastern United States lost genetic diversity to both a bottleneck and a founder effect. The wolves were eradicated in the wild, and just a handful of individuals remained in captivity. These captive wolves became the founders of a recovery effort. Breeding has increased wolf numbers and enabled reintroduction to the wild, but the genetic diversity of the population is very low.

REMEMBER

Populations with lower genetic diversity are less resilient to environmental change. When selection pressures change, individuals with the most beneficial combination of alleles are more likely to survive and reproduce, continuing the population. If the individuals in a population are genetically very similar, there are fewer allele combinations and a lower chance of survival.

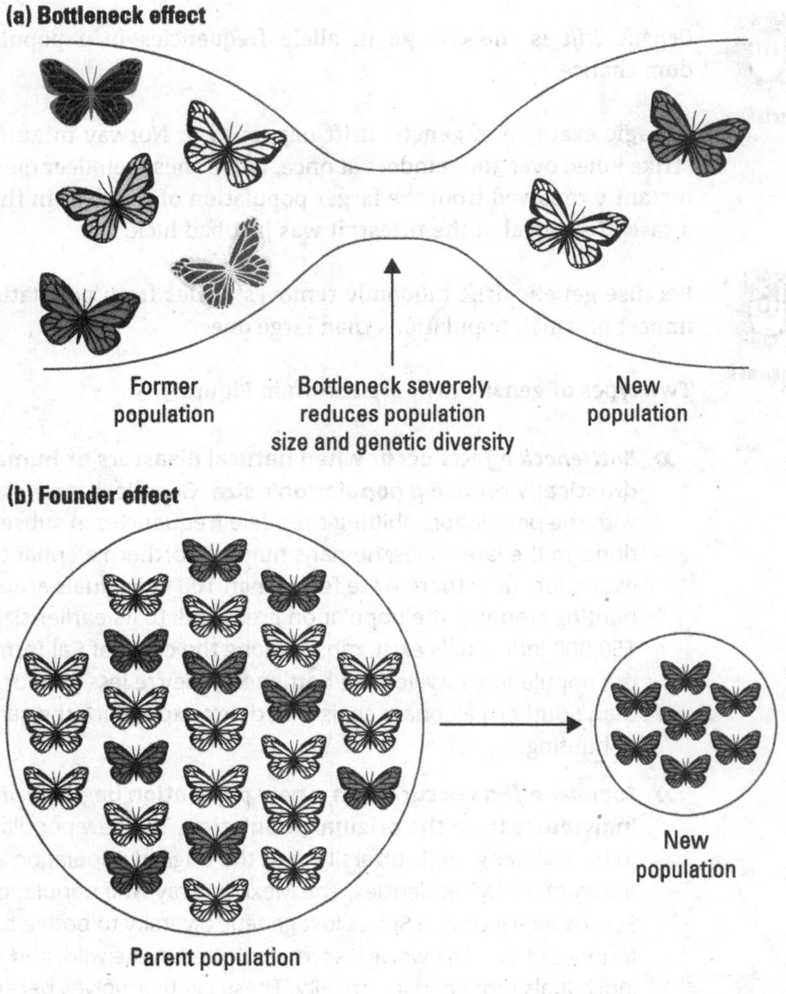

FIGURE 3-5: Genetic drift.

Sometimes, ancestral populations evolve over a relatively short period of time into groups of related organisms that show great diversity in form and function.

REMEMBER

Adaptive radiation occurs when a lineage diversifies rapidly, producing many new species.

Adaptive radiation seems to result from the ability of individuals in the ancestral population to take advantage of unique opportunities:

- » **Lack of competition.** Mass extinction or the formation of a new volcanic island makes new resources available to existing populations. Darwin observed many species of finches in the Galapagos Islands that weren't present on mainland Ecuador. When ancestral mainland finches arrived on the island, very few types of birds lived there. Without competition, the finches were able to take advantage of many different types of food, as shown in Figure 3-6. Over time, different populations adapted to efficiently utilize the different food sources, eventually becoming distinct species.
- » **Evolution of a key adaptation.** The development of an ability that makes new niches available to an ancestral population can foster adaptive radiation. (A *niche* is the specific lifestyle of an organism, which is discussed further in Chapter 7.) Plant-eating beetles in the group Phytophaga rapidly diversified right around the time they acquired genes that allow them to digest plant cell walls. The genes are only found in a couple of groups of beetles and seem to have come from bacteria and fungi. These genes allowed the beetles to make proteins they needed to bore into woody tissues, allowing them access to new sources of food.
- » **Specialization.** An ancestral population that occupies a particular niche may subdivide that niche, leading to adaptive radiation as groups specialize on each resource. Cichlid fish focus on different aspects of their niche, leading to speciation and generating a wide variety of body shapes, color patterns, and behavior.

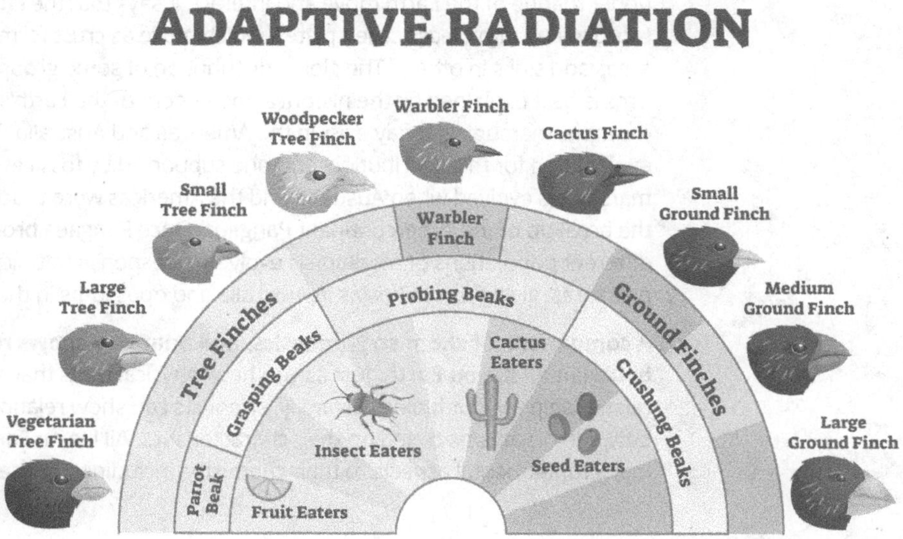

FIGURE 3-6: Adaptive radiation.

VectorMine/Adobe Stock Photos

Looking for Evidence of Evolution

Scientists may debate the details of how and when different evolutionary events occurred, but they agree that evolution is an important force in the history of life on Earth. Evolution can help people understand many aspects of the world around us, from the evolution of antibiotic-resistant bacteria or new strains of viruses, to predictions of the impacts of climate change on populations. Although evolution is central to a scientific understanding of the natural world, the idea is sometimes seen as controversial either because it's misunderstood or because it seems to conflict with people's personal beliefs.

REMEMBER

The theory of evolution says that all life on Earth is related and evolved from a common ancestor.

In science, a *theory* is an idea that's supported by multiple lines of evidence. Here are some examples of the evidence that supports the theory of evolution:

- **The fossil record shows evidence of organisms that are now extinct, including organisms that bridge the gap between groups.** From the fossil record, we know that living things today are not the same as living things from long ago, so we can infer that living things change over time. Fossils like the feathered, bird-like reptile *Archaeopteryx* support the idea that a new group of organisms (birds) can evolve from an ancestral group (reptiles).
- **Biogeography shows evidence of change in the distribution of closely related organisms.** *Plate tectonics* is the explanation of how the crust and upper mantle of the Earth move and interact. It says that the crust of the Earth is broken up into pieces called plates, which move as crust forms in some areas and sinks in others. The global distribution of some groups of organisms is best explained by the historical movement of the Earth's plates. For example, marsupials today exist in the Americas and Australia. The best explanation for this distribution, and one supported by fossil evidence, is that marsupials evolved when Australia and the Americas were touching, before the breakup of the supercontinent Pangaea. Once Pangaea broke apart, the different populations of marsupials evolved in response to unique selection pressures, giving rise to koalas in Australia and opossums in the Americas.
- **A comparison of chemistry, genetics, and anatomy shows relationships between all life on Earth.** Just as you have physical traits that show your relationship to your biological family, scientists can show relationships between organisms based on their characteristics. All life on Earth shares many fundamental aspects to their chemistry, including the use of DNA as the

genetic code. When scientists compare specific gene sequences, they can see similarities across broad groups of organisms. Similarities in physical structures, like the bones shown in Figure 3-7, also show relationships. When scientists examine living things today and build family trees based on similarities, the trees that result show that all life is related, and they show how some groups of organisms today connect to groups from the past.

» **Scientists can observe evolution happening today.** Evolution in real time is easiest to see in rapidly growing organisms like bacteria, where we can see the evolution of antibiotic resistance. But scientists have also seen small changes occurring in other types of living things, like the modern-day populations of finches on the Galapagos Islands. Two evolutionary biologists, Peter and Rosemary Grant, went to the Galapagos to study Darwin's finches. They took measurements of the finches' characteristics, like size and beak length, and were amazed to see that they could detect changes from one generation to the next. They stayed and continued to study the finches for 40 years, eventually witnessing the evolution of a new species of finch.

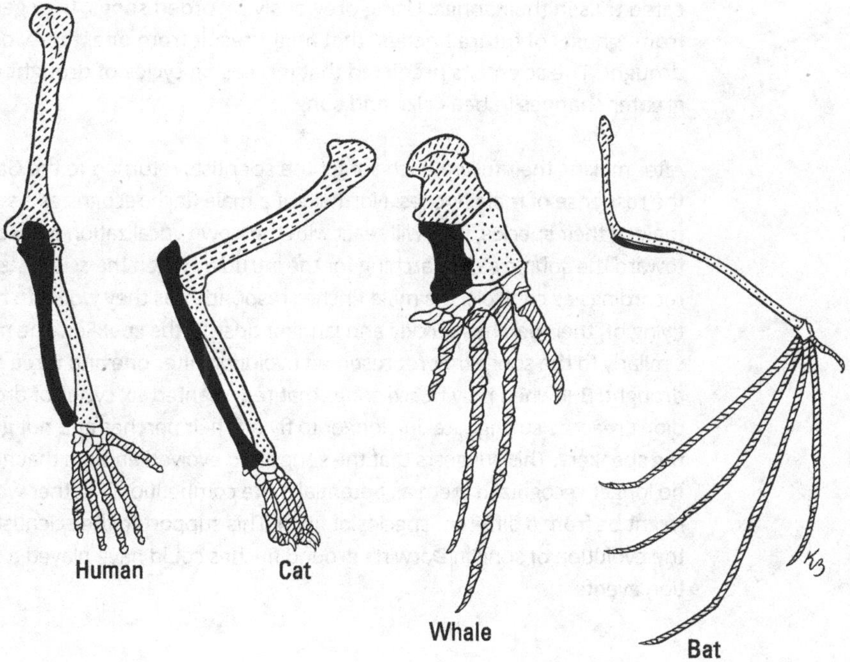

FIGURE 3-7: Comparative anatomy of the bones in the front limbs of humans, cats, whales, and bats.

FINCHES FROM THE FUTURE

Darwin's finches from the Galapagos Islands are a famous example of speciation through adaptive radiation. After the founder finches landed on the islands, their descendants specialized in different food sources. As finches occupied different niches, disruptive selection led to the evolution of new species with distinct beak types (shown in Figure 3-6).

Recently, scientists at the University of Massachusetts in Amherst wanted to test whether changes in bird song could have played a role in these speciation events. Evolutionary biologist Jeffrey Podos knew that changes in beak structure lead to changes in song. Finches use their songs to attract mates, so Podos wondered if changing beaks could contribute to reproductive isolation of different populations. Stopping gene flow between populations would increase the chance of speciation.

First, Podos and his student Katie Schroeder modeled how the beaks of Darwin's medium ground finches would change in response to drought. They knew from previous studies that drought led to changes in seed size, which, in turn, led to bird populations with increased beak size. Birds with larger beaks can't perform complicated trills in their songs. Using previously recorded songs, they generated songs from "ghosts of future finches" that might result from one, three, or six periods of drought. The scientists predicted that increasing cycles of drought would lead to greater changes in beak size and song.

After making their future finch songs, the scientists returned to the Galapagos to test the response of male finches. Normally, if a male finch recognizes a song from another male of their species, they will react with their own vocalization and sometimes fly toward the source as if searching for the intruder. When the scientists played unaltered recordings as controls, the male finches responded as they would to hearing a live bird, flying off their perches quickly and landing close to the speaker. The males responded similarly to the songs that represented evolution after one and three periods of drought. But when they heard songs that represented six cycles of drought, the males didn't react as strongly, taking longer to fly off their perches and not going as close to the speakers. This suggests that the songs had evolved enough that ground finch males no longer recognized them as potential mate competition, in other words, that they might be from a different species of finch. This supported the scientists' hypothesis that the evolution of song in Darwin's ground finches could have played a role in speciation events.

IN THIS CHAPTER

» Exploring some fundamentals of life on Earth

» Surviving in the sea

» Taking on the challenges of life on land

Chapter 4

Adapting to Different Environments

From space, Earth is a beautiful blue sphere due to the oceans that cover 70 percent of its surface. Early life evolved in these oceans, developing adaptations to the physical constraints of a watery environment. Then, after a billion years of life in the water, microbial life moved onto the land, followed later by plants and then animals. These organisms had to adapt to a completely different reality.

Today, life is found in every environment on Earth, from the saltiest waters to the deepest, darkest caves. Although external environments vary widely, living things have evolved ways of getting what they need and maintaining their internal conditions within a range that allows survival.

REMEMBER

Homeostasis is a self-regulating process by which biological systems maintain stability while adjusting to changing external conditions.

In this chapter, you'll first investigate some of the most important fundamental conditions that define and challenge life, then you'll take a look at some of the adaptations that have enabled life to be successful in the water and on land.

Life, the Environment, and Everything

Living things are woven into the environment that surrounds them. Plants push their roots through the soil, fish swim in the sea, and people breathe in the atmosphere. Every living thing takes what it needs to survive and releases waste products back out again. Matter and energy constantly flow back and forth between the environment and organisms, but the living are distinctly different from the non-living.

REMEMBER

Biotic factors are living things within an ecosystem, including plants, animals, and bacteria, while *abiotic factors* are non-living components like water, soil, and atmosphere.

Defining the boundary of life

All living things are made of cells, and cells are separated from the environment by a boundary called the *plasma membrane*. The plasma membrane controls what enters and exits cells, helping to maintain conditions necessary for life.

REMEMBER

Plasma membranes are *semipermeable*, meaning that some things can cross the membrane without assistance, but others can't.

The plasma membrane, shown in Figure 4-1, is made of two layers of phospholipids, forming a phospholipid bilayer with the hydrophilic heads of the phospholipids pointing outward and the hydrophobic tails of the phospholipids sandwiched in the middle.

REMEMBER

Hydrophobic molecules repel water, while *hydrophilic* molecules are attracted to water.

Living things move what they need into cells and move waste products out. Whether or not a molecule will cross a membrane, and which way it will go, depends upon three things:

- **The chemical structure of the molecule, particularly its size and charge.** Small, hydrophobic molecules can cross the lipids in the plasma membrane all by themselves. As molecules get larger, or if they're hydrophilic, they can cross the membrane only if a transport protein in the membrane helps them across.

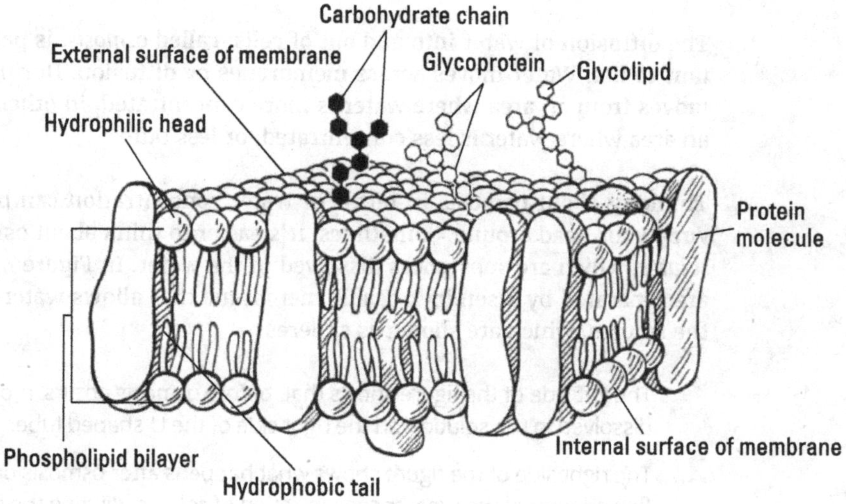

FIGURE 4-1: The fluid mosaic model of plasma membranes.

- **The concentration of the molecule on either side of the membrane.** If molecules are allowed to move freely, they will move from areas where they're more concentrated to areas where they are less concentrated — in other words, they'll move until they're evenly spread out, or in *equilibrium,* over an area. This type of movement from an area of higher concentration to an area of lower concentration, called *diffusion*, is shown in Figure 4-2.
- **Whether or not the membrane contains a transport protein for the molecule.** Although some small hydrophobic molecules can cross membranes by themselves, most molecules need the help of transport proteins. If a plasma membrane doesn't have a protein to transport a particular molecule, it won't be able to cross in or out of the cell.

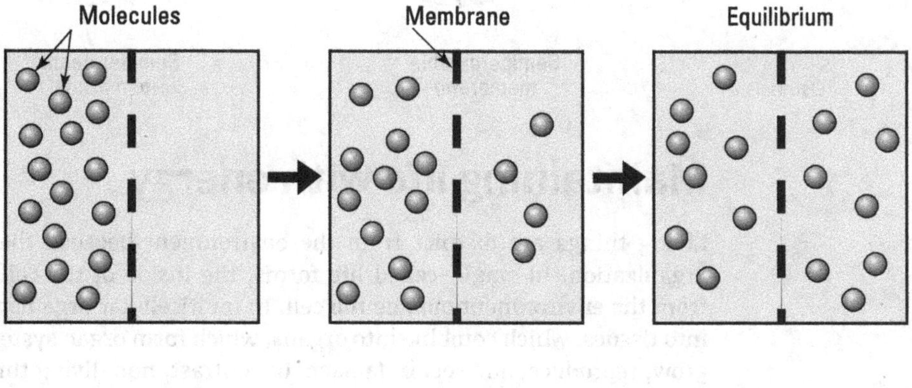

FIGURE 4-2: Diffusion.

The diffusion of water into and out of cells, called *osmosis*, is particularly important to life. Water moves across membranes by diffusion. During osmosis, water moves from an area where water is more concentrated, in other words, purer, to an area where water is less concentrated, or less pure.

Thinking about osmosis in terms of water concentration can be a little hard to wrap your head around. Sometimes, it's easier to think about osmosis in terms of *solutes*, which are substances dissolved in the water. In Figure 4-3, two solutions are separated by a semipermeable membrane that allows water to cross, but not the solutes, which are shown as spheres:

1. The left side of the figure shows that, before osmosis occurs, more solutes are dissolved in the solution on the right side of the U-shaped tube.
2. The right side of the figure shows what happens after osmosis occurs. Water flowed toward the greater concentration of solutes, diluting the solution on the right until it was equal in concentration to that on the left.

TIP

One way to remember which way water will move by osmosis is to remember the phrase "salt sucks." Salt is a solute, and water moves into solutions with more solutes. So, a solution with more "salt" (solutes) will "suck" water toward it.

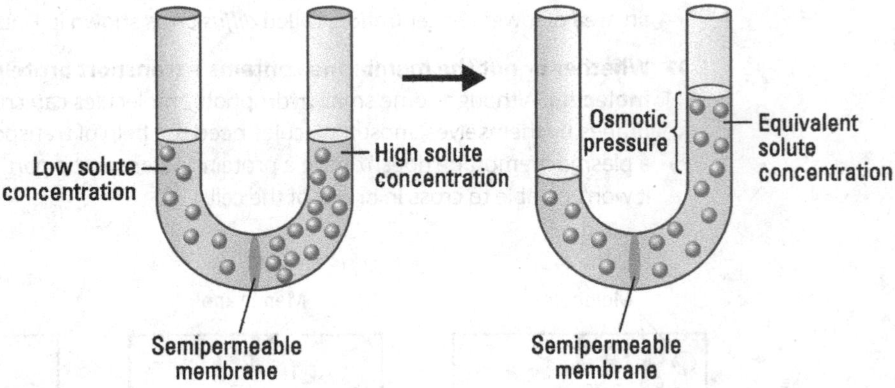

FIGURE 4-3: Osmosis.

Maintaining life with energy

Living things are distinct from the environment because they maintain their organization. In single-celled life forms, the inside of the cell is very different from the environment outside the cell. In multicellular organisms, cells organize into tissues, which combine into organs, which form organ systems. Living things grow, reproduce, and repair damage. In contrast, non-living things always move toward greater randomness, or *entropy*. Rocks weather into minerals, water flows downhill, and heat disperses.

To fight entropy and stay organized in a way that supports life, living things need a constant supply of energy.

REMEMBER

Energy is something that allows work to be done.

Organisms use energy to do work like building molecules, transporting substances, repairing damage, moving, and reproducing. Without energy, living things can't maintain themselves, and they die.

Most of the energy for life on Earth traces back to the Sun. Photosynthetic organisms capture light energy, transforming it into chemical energy stored in food molecules. Organisms then use these food molecules to supply the energy they need to maintain life.

The following chemical reaction summarizes the process of photosynthesis:

$$\underset{\text{carbon dioxide}}{6CO_2} + \underset{\text{water}}{6H_2O} + \text{light energy} \rightarrow \underset{\text{sugar}}{C_6H_{12}O_6} + \underset{\text{oxygen}}{6O_2}$$

REMEMBER

The overall purpose of photosynthesis is to combine matter and energy into food in the form of carbohydrates. For more details on the process itself, head to Chapter 10.

The light energy needed for photosynthesis is the result of nuclear fusion reactions inside the Sun. These reactions produce a tremendous amount of energy in the form of *electromagnetic radiation* that includes visible light, X-rays, microwaves, and ultraviolet radiation (see Figure 4-4). Plants and other photosynthetic organisms capture electromagnetic radiation within the visible light band for photosynthesis.

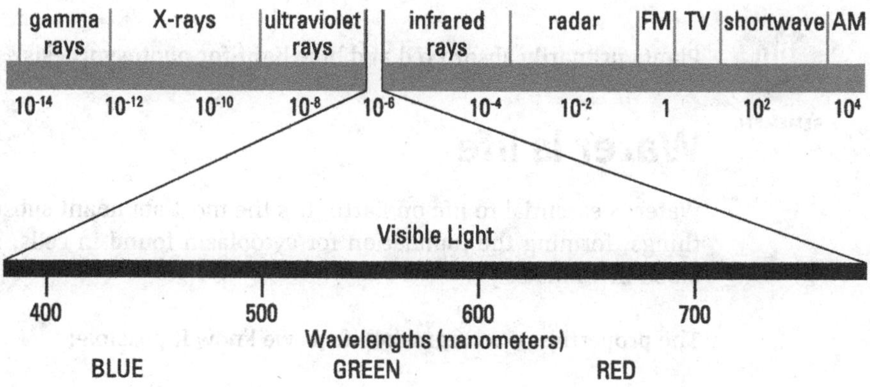

FIGURE 4-4: The electromagnetic spectrum.

Plants use *pigments* to absorb light. The most abundant pigment in plants is *chlorophyll*, which looks green because it absorbs blue and red light, but not green (see Figure 4-5). Green light hits the leaves of the plant and bounces off again, traveling to your eye and causing your brain to interpret the color as green. Photosynthetic organisms also have *accessory pigments*, like those shown in Figure 4-5, that absorb different colors of light, increasing the total amount of light that can be harnessed for photosynthesis. Different types of plants and algae are differently colored because of the types of accessory pigments they have.

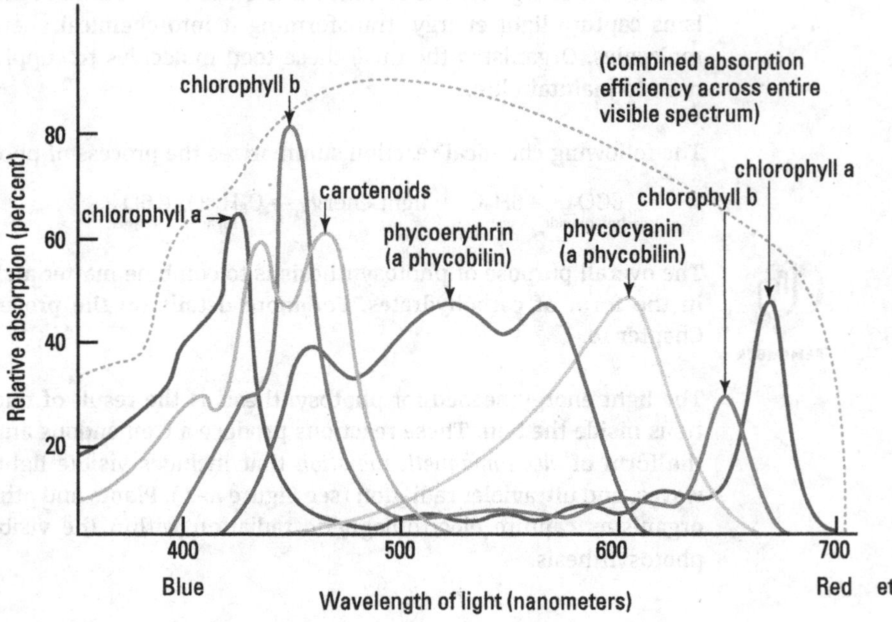

FIGURE 4-5: Absorption spectra for photosynthetic pigments.

REMEMBER

Plants primarily absorb red and blue light for photosynthesis.

Water is life

Water is essential to life on Earth. It's the most abundant substance inside living things, forming the foundation for cytoplasm found in cells, sap in plants, and blood in animals.

The properties of water make life as we know it possible:

» **Water is polar.** Water molecules have partial positive charges on the hydrogen atoms and partial negative charges on the oxygen atoms. These partial charges make water attractive to other charged molecules, including other water molecules:

- Water molecules have strong *cohesive* forces due to their ability to attract each other. This electrical stickiness between water molecules creates *surface tension*, the resistance that allows water bugs to walk across a pond.
- Water has *adhesive* properties that allow it to stick to substances other than itself. Adhesion helps water cling to the sides of plant cells as water moves upward against gravity.

» **Water is an excellent solvent.** Water sticks to many charged atoms and molecules, causing them to *dissolve* or disperse their components into the water. As water moves through the hydrologic cycle, shown in Figure 4-6, it picks up atmospheric gases and minerals, carrying them to the ocean as it moves in rivers and streams. As pure water evaporates from the surface of the ocean, minerals and salts are left behind and concentrated.

» **Water has high heat capacity.** Water resists changes in temperature, remaining liquid over a broad range. It takes a lot more energy to change the temperature of water one degree Celsius than it does air. Because of this, water warms up and cools down more slowly than the air, helping to regulate temperature in the environment. Thus, the temperature of the water of a pond might remain relatively constant from day to night, regardless of the changing atmospheric temperature.

» **Water is less dense as a solid than as a liquid.** As water freezes, the molecules form a crystalline structure that spaces the molecules further apart than in liquid water. This means that ice is less dense than liquid water, which is why it floats. This property keeps ponds, lakes, and oceans from freezing solid and allows life to continue to live under the icy surface in winter.

The concentration of hydrogen ions (H^+) dissolved in water determines its *acidity*, which is commonly measured according to the pH scale (Figure 4-7). Because hydrogen ions are charged particles, they can interfere with the structure of biological molecules like enzymes, causing damage to living things. Acidity in the ocean causes corals to release their algal partners, leading to coral bleaching and the death of coral reefs.

The pH scale ranges from 0 to 14. Most living things prefer a pH of about seven, which is neutral. In solutions that are pH neutral, the number of positively charged hydrogen ions (H^+) is balanced by the number of negatively charged hydroxyl (OH^-) ions.

REMEMBER

Acidic solutions have a higher concentration of H^+ ions, but this results in a lower pH. *Basic*, or *alkaline*, solutions have a lower concentration of H^+, which results in a higher pH.

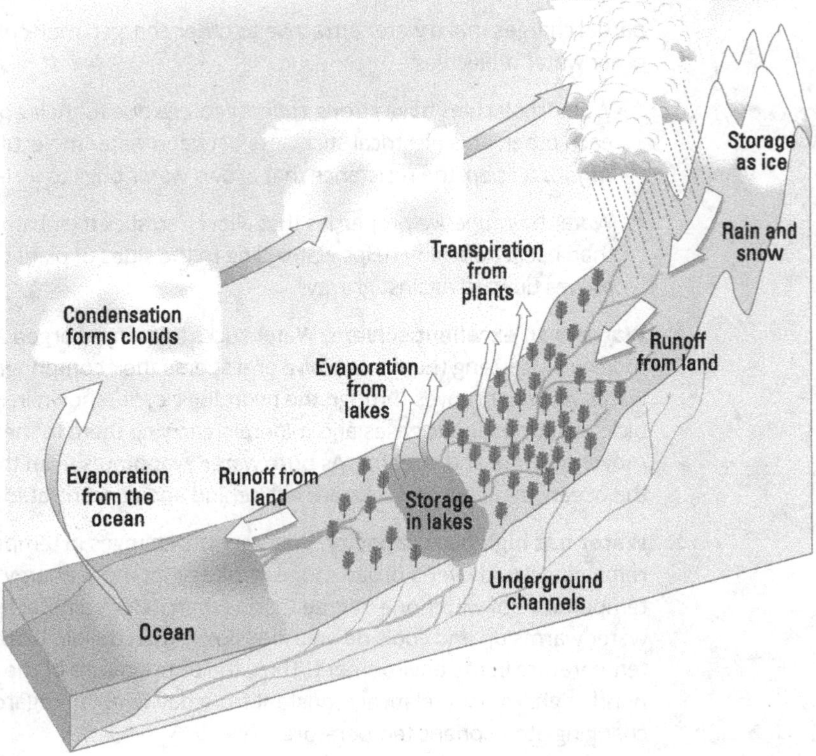

FIGURE 4-6: The water cycle.

TIP The letter A comes before the letter B in the alphabet, so acids come before (are lower than) bases on the pH scale.

In the early 1900s, scientists came up with the *pH scale*, a system of classifying how acidic or basic a solution is. The term *pH* symbolizes the hydrogen ion concentration in a solution (for example, what proportion of a solution contains hydrogen ions).

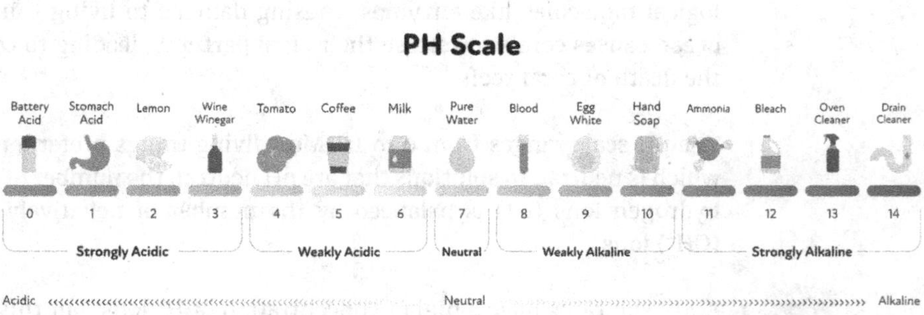

FIGURE 4-7: The pH scale.

tasty_cat/Adobe Stock Photos

Evolving in Water

Life on Earth evolved in the oceans for more than a billion years before the first life forms made their way onto land. Today, the oceans cover 70 percent of the Earth's surface, with an average depth of 3,682 meters (12,080 feet). Life exists throughout this huge area, but much of it is microscopic. An often very large exception to this is ocean animals. The largest animal on Earth is the Antarctic blue whale, weighing up to 400,000 pounds and reaching up to 98 feet in length.

Swimming and floating

Water is dense, providing support for organisms, but also viscous, which means it resists movement through it. Organisms that live in saltwater or freshwater need adaptations to help them float or swim:

- Swimming organisms often have streamlined body shapes that reduce drag.
- Many small organisms have tendrils or other long appendages that move slowly through the viscous water, helping the organisms stay afloat.
- Organisms can have chambers or bladders that contain air, acting as a flotation device:
  - Some fish have gas-filled swim bladders that adjust in order to match the fish's body density to that of the ocean.
  - Plants like kelp have gas-filled bladders near their leaves to keep the leaves floating in areas where light is available. Other water-adapted plants have air-filled tissues, called aerenchyma, that help them float (see Figure 4-8 for an example of aerenchyma).
- Freshwater aquatic plants, like those shown in Figure 4-8, may have flat leaves that float at the water's surface or small, rounded bodies that float easily.

Dealing with salt (or the lack of it)

Organisms that live in the ocean are surrounded by saltwater. If you remember the phrase "salt sucks," then you might realize that this presents a problem. The concentration of salt outside the bodies of marine organisms is higher than the concentration of solutes inside their bodies. This means that water could move by osmosis out of their bodies and into the sea. This is also the reason you should never drink seawater if you're lost at sea: you will actually draw water out of your tissues into your digestive tract. The result of losing water is dehydration and potential death.

FIGURE 4-8: Adaptations of aquatic plants.

REMEMBER

Osmoregulation refers to the mechanisms that organisms use to maintain the proper salt balance within their body fluids.

Marine organisms have developed three main strategies for coping with life in saltwater:

» **Conform to the salt:** In *osmoconformers,* the concentration of solutes in their extracellular fluids matches that of the environment.

- Most marine crustaceans are very permeable to water, allowing it to pass from the environment into their fluids. To protect against water loss from inside their cells due to osmosis, they adjust the concentration of small solutes in the cytoplasm of the cells.
- Some sharks and rays retain urea in their bloodstream instead of excreting it in urine. The urea balances their internal fluids with the salt in seawater, preventing water loss due to osmosis.

» **Push out excess salt:** Marine fish drink seawater to replace their lost water, using their kidneys to excrete the excess salt back out into the environment in urine. As seawater passes over their gills, specialized cells on the gills pump out the excess salt.

» **Avoid as much salt as possible:** Marine mammals like whales and dolphins don't drink seawater; instead, they get the water they need from their food. Their kidneys concentrate salt, producing very salty urine that gets released back into the ocean.

Organisms that live in freshwater have the opposite problem from marine organisms. Their cells have a greater solute concentration than the environment. Water moves from the environment into their cells, and solutes may diffuse out into the environment. To prevent their cells from exploding or losing their solutes, organisms have adaptations to cope with life in an environment with too little salt:

» **Have a cell wall:** The cells of plants and algae are surrounded by a semi-rigid cell wall. As water flows into their cells, the cell wall exerts pressure on the cell, countering the flow from osmosis.

» **Eliminate excess water:** As water flows in, some organisms move it back out again:

- The mouths and gills of freshwater fish are very permeable to water, which constantly moves into their bodies by osmosis. The kidneys filter the fluid waste of the fish, collecting the solutes and returning them to the blood, then releasing the excess water into the urine. The gills also absorb ions from water and move them into the blood.
- Many protozoa that lack cell walls have an organelle called a *contractile vacuole*. As water flows into the cells of the protozoans, it's channeled into the contractile vacuole. The cell uses energy to periodically contract the vacuole, squirting the water back out into the environment.

In cells with cell walls, the movement of water is affected both by the presence of solutes and the pressure inside the cell. The net result of these two factors can be expressed in terms of *water potential* (Ψ), which is the sum of the pressure due to solutes, called *solute potential* or *osmotic potential* (Ψ_S), plus the internal pressure of the cell (Ψ_P), or:

$$\Psi = \Psi_S + \Psi_P$$

REMEMBER

Water always moves from an area of higher water potential to an area of lower water potential.

The solute potential of pure water is zero. The addition of solutes decreases the solute potential, making the overall water potential more negative. This is another way of thinking about the idea that "salt sucks," or that water moves toward the greater concentration of solutes.

As water moves into a walled cell, the pressure potential increases. As pressure potential increases, water potential increases, which potentially slows down the movement of water by osmosis. In other words, the negative solute potential of the cytoplasm of a walled cell will attract water by osmosis, gradually increasing pressure until the two factors balance each other and the overall water potential becomes zero.

Exchanging gases

Gases like carbon dioxide (CO_2) and oxygen (O_2) diffuse from the atmosphere into water. These gases diffuse more slowly through water than they do through air, which makes them less available to organisms that live in water than to those that live on land.

Most carbon dioxide molecules change form when they dissolve in water, becoming carbonic acid (H_2CO_3). Depending upon the pH of the water, carbonic acid releases either one or two hydrogen ions (H^+) into the water, forming either bicarbonate (HCO_3^-) or carbonate ions (CO_3^{2-}). Bicarbonate is the most common form in most salt- and freshwater environments, which typically have a pH between six and nine. Aquatic plants and algae can use either carbon dioxide or bicarbonate for photosynthesis. Because these ions diffuse slowly through water, the availability of carbon can be a limiting factor for photosynthesis in aquatic environments.

Oxygen isn't very soluble in water, so it's present at lower concentrations in aquatic environments than it is on land. Plus, oxygen production via photosynthesis is limited in aquatic systems to the areas where light is available.

REMEMBER

The *photic zone* is the area in aquatic environments that has enough light for photosynthesis. In the ocean, the photic zone may reach as deep as 200 meters.

Many aquatic organisms solve the problem of low oxygen availability by having lower metabolic rates or by surviving for longer periods on metabolic pathways like fermentation that don't require oxygen. Oxygen may diffuse directly through the outer layer of an organism or be absorbed from the water through gills. Fish have evolved an extremely efficient mechanism, called *countercurrent flow* (Figure 4-9), to extract the most oxygen they can out of water:

- **Water flows from the environment over the gills in one direction.** As the water moves over the gills, oxygen diffuses into the blood of the fish. Thus, the concentration of oxygen in the water decreases as it moves over the gills.
- **Blood flows over the gills in the opposite direction to the water.** As the blood flows through the gill filaments, the concentration of oxygen in the blood increases.
- **At every position along the gills, the oxygen concentration in water is higher than that in the blood, so that oxygen diffuses from the water to the blood.** Even when the concentration of oxygen in the water is low, it passes next to blood with an even lower concentration, allowing oxygen to continue to move into the blood.

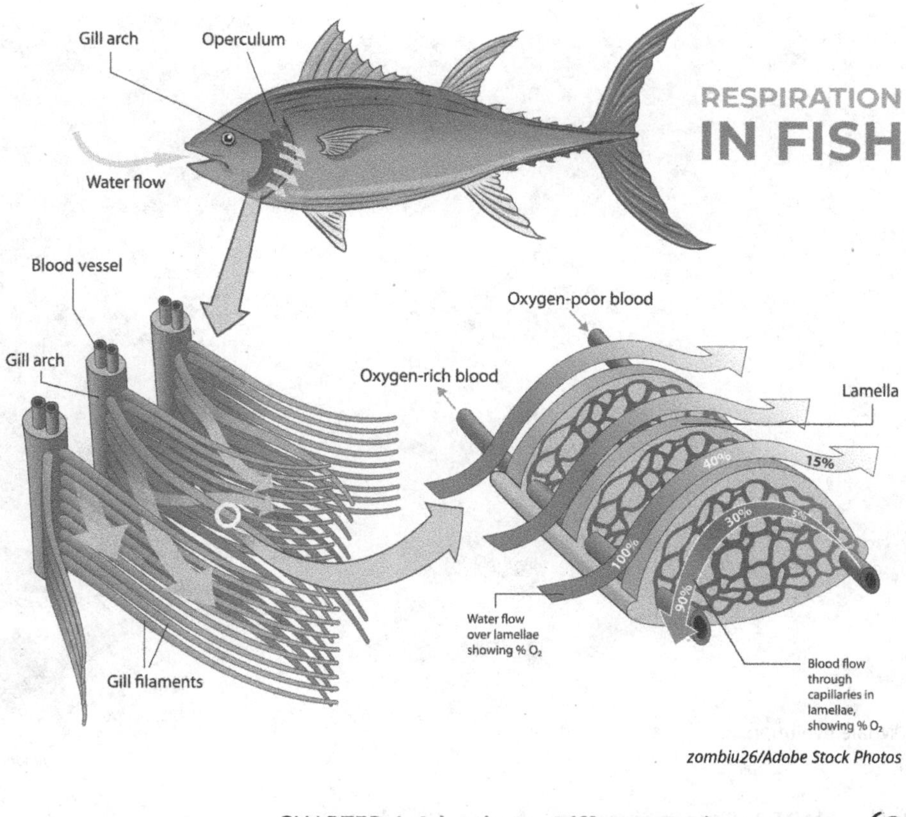

FIGURE 4-9: Countercurrent flow in fish gills.

zombiu26/Adobe Stock Photos

Marine mammals evolved from land animals, retaining their ancestral habit of breathing air into lungs. Figure 4-10 shows the fossil record of the ancestors of modern whales, revealing how they adapted to their new home over time (for more on the evidence that supports the evolution of whales, see the sidebar "A whale of a tale"). As whales, dolphins, and porpoises evolved, their nostrils moved higher onto their heads, becoming a blowhole that made it easier to breathe air at the surface of the water.

The low concentration of oxygen, combined with its slow diffusion rates in water, causes some deep-water environments to lack oxygen entirely. Life in these *anaerobic* or *anoxic* environments must either be able to survive without oxygen or must carry its oxygen with it:

- Many microbes have metabolic pathways that run without oxygen. Some of these form the basis of food webs around hydrothermal vents on the ocean floor.
- Deep-diving whales can survive on a single breath for up to an hour, which allows them to swim through low-oxygen environments.

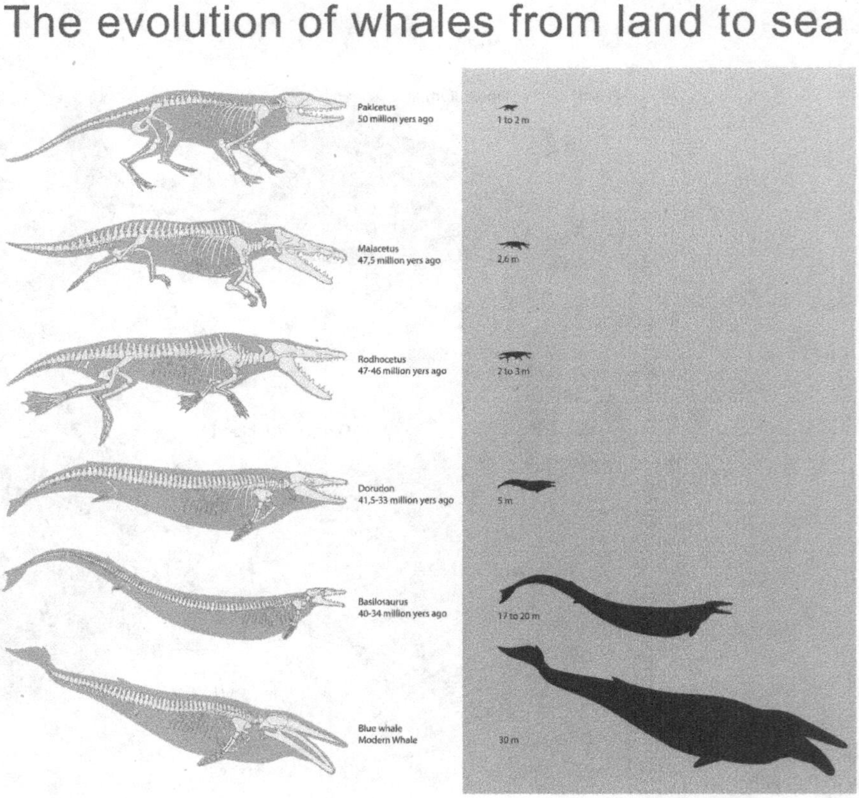

FIGURE 4-10: Whale evolution.

Aldona/Adobe Stock Photos

A WHALE OF A TALE

Whales move through the ocean gracefully, using their flukes to propel their sleek bodies through the water. When they need oxygen, they rise to the surface, blowing out water and then taking in air with the blowholes on the top of their heads. They have blubber to insulate them from the cold, and they can tolerate the pressure of deep ocean dives. They're obviously well-adapted to their environment, with many superficial similarities to fish. These similarities, combined with the fact that whales live in the ocean, led scientists to classify whales as fish for almost 2,000 years despite the fact that they knew whales produce milk, give birth to live young, and even produce a few hairs.

You probably know that whales have these traits because they are mammals. What you might not realize is that the earliest mammals were land-dwelling creatures that walked around on four legs. When scientists finally put these two pieces of information together, they realized that whales evolved from mammals that moved from the land to the ocean, losing their legs along the way.

To figure out which group of mammals was most related to whales, scientists first turned to the fossil record, carefully examining fossils of primitive whale-like creatures that lived between 45 and 25 million years ago, like *Dorudon* and *Basilosaurus* in Figure 4-10. These fossils had tiny leg bones that correspond to the structure of land animal leg bones, and also had very distinctive teeth. When scientists looked for more fossils with similar teeth, they found fossils that had many similarities with the primitive whales, but that also had legs and even hooves (see Figure 4-10). These fossil animals belong to an extinct group of meat-eating, hooved mammals called Mesonychids.

Mesonychids might be extinct, but two living groups of hooved mammals, or *ungulates*, are alive today. Even-toed ungulates, called *artiodactyls*, include hippos, pigs, giraffes, and camels. Odd-toed ungulates, called *perissodactyls*, include horses, zebras, tapirs, and rhinos. To decide which of these groups was most closely related to whales, scientists took a very close look at the ankle bones of fossil whales. Based on the number of ridges on the ankle bones, whales are more closely related to artiodactyls than they are to perissodactyls.

The last piece of the puzzle was to figure out which artiodactyls are whales' closest relatives. For this, scientists compared the actual DNA codes of several artiodactyl species with those of whales and found that hippos are the closest living relatives of whales! Scientists now think that both hippos and whales evolved from a common ancestor, and many scientists group them in a subgroup of the artiodactyls called Whippomorpha (because whale + hippo = whippo).

Learning to Live on Land

Organisms that moved onto the land faced a whole new set of challenges, most significantly the need for water. Animals can move to find water, but plants are restricted to growing in one place. The loss of a watery environment also results in more extreme temperature shifts and exposure to greater levels of solar radiation. These challenges are significant, but life found a way to adapt.

Getting water and nutrients

Plants have evolved sophisticated systems for acquiring and storing water, as well as preventing water loss.

Taking in water and nutrients with roots

Water enters plants through the roots, traveling by osmosis from the soil into root cells. Root cells contain solutes and have a lower water potential than pure water. Water in the soil, however, clings to the soil particles, forming multiple layers of water shells around the particles. The water shells farthest from the soil particles have the highest water potentials, so water flows most easily from these shells into the root cells. As the shells get closer to the particles, they have lower water potentials and become less accessible to plants. So, roots typically can't take all the water from the soil.

REMEMBER

Soil contains a mix of particles of clay, silt, sand, and organic material. Clay particles are the smallest, so soils with high levels of clay have more surface area, allowing them to hold more water. Sand particles are the largest, producing soils with less surface area through which water drains more quickly.

As roots take up water, they also take up *mineral nutrients* for plant growth. As rocks in the soil dissolve, they release ions into the soil. It's not always easy for plants to get these minerals because positively charged ions, called *cations,* cling to the negatively charged surfaces of clay particles in the soil. Plants can loosen minerals up by releasing cations, such as hydrogen ions (H^+), which are also attracted to the clay and can bump off some of the other cations, making them available for absorption by the plant.

Moving water with transpiration

Plants transport water through their *xylem,* which is mostly made up of dead cells whose strong secondary cell walls form tubes throughout the plant, much like pipes form the plumbing in your home.

Water moves through plants by a combination of two factors:

- The adhesive and cohesive properties of water
- The pull of water evaporation from the leaves

REMEMBER

The evaporation of water from the leaves is called *transpiration*. The pulling force of transpiration is the opposite of the pushing force of pressure, so scientists refer to it as *negative pressure* or *tension*.

Water evaporates from leaves through tiny pores called *stomates* on the underside of plant leaves. As transpiration occurs, it pulls water upward through the plant, as shown in Figure 4-11.

- Water evaporates from air spaces in the leaf into the atmosphere.
- Adhesive forces between the water molecules and the cell walls of the xylem draw the water column up the xylem. The water column will rise until the downward pull of gravity balances the adhesive forces. Xylem tubes are very narrow, so lots of contact occurs between the water and the cell walls, helping the upward movement of the water column.
- Cohesive forces between the water molecules hold the column of water together.

TIP

Water moving through a plant is very similar to what happens if you drink water through a straw. You provide negative pressure with your mouth, just like transpiration. As you suck on the straw, water clings to the edges of the straw through adhesion and sticks together in a column because of cohesion. This is exactly what happens inside the tubes of xylem in a plant.

REMEMBER

The explanation for how water moves through plants is called the *cohesion-tension theory*.

Preventing water loss

The surfaces of organisms that live on land are constantly exposed to the atmosphere, which has a very negative water potential, causing organisms to constantly lose water. Organisms can counter this by taking in more water, but this can be a challenge for those that live in dry environments like deserts. To survive, these organisms developed adaptations to conserve as much water as possible.

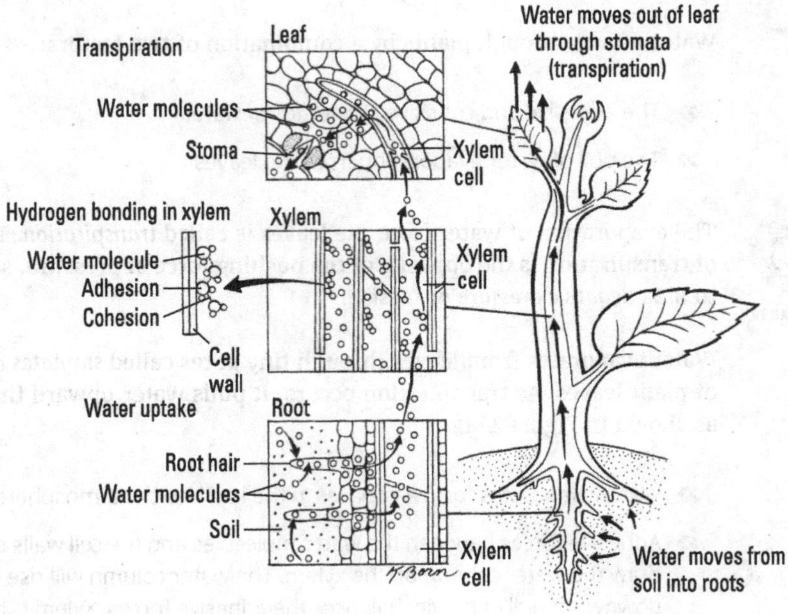

FIGURE 4-11: The movement of water through a plant.

Desert plants

Desert plants face water scarcity, temperature extremes (hot during the day, but cold at night), and high amounts of solar radiation. These conditions put strong selection pressures on plants, resulting in the familiar adaptations of desert plants like those shown in Figure 4-12:

» **Reduced or highly modified leaves:** Thin, flat plant leaves may be ideal for catching sunlight in a moist area, but they are a liability in the desert. Those thin, flat surfaces would be exposed to lots of dry air, causing water to evaporate rapidly out of the leaves.

- In many desert plants, such as the barrel cactus, leaves have evolved into spines, preventing water loss and providing protection from grazers. These plants have switched the main site of photosynthesis to the stem.
- Other desert plants, such as the agave, have evolved tough, thick leaves that have a heavy wax coating to prevent water loss.
- Some plants, such as the desert yam, produce delicate, thin leaves that have the potential for high water loss. The plants get around this by only growing leaves during the rainy season. The rest of the year, the leaves die back, leaving only the plants' tough, protective base.

» **Organs for water gathering and storage:** Many desert plants are *succulents,* plants that store water in fleshy tissue. All four desert plants shown in Figure 4-12 have ways of dealing with the desert water shortage. Cacti store water in their stems, which have pleats to allow expansion when water is available. Agave plants have long root systems that can find water under the desert ground. The desert yam stores water in its tough base, and stone plants have succulent stems.

a. Spiny leaves and succulent, photosynthetic stems (barrel cactus)

b. Succulent leaves (agave)

c. Seasonal leaf production (desert yam)

d. Reduced exposure to light (stone plants)

FIGURE 4-12: Common strategies plants use to survive desert conditions.

- **Protection from overexposure:** Too much light or exposure to dry air can damage a plant, so desert plants have evolved structures and strategies to protect themselves from these conditions. The long, white, hair-like spines that grow on some cacti form a surface barrier over the stem, slowing down air movement and reducing transpiration. Because the spines are white, they also help reflect excess sunlight off the stem, providing protection from overexposure. Some plants, such as the stone plant shown in Figure 4-12, embed themselves into the desert floor, allowing just the upper surface of the stem to come into contact with the environment. Clear spots on the surface of the stem act as windows that allow sunlight to enter the plant.

Desert plants face a particularly complex puzzle when it comes to photosynthesis. They need carbon dioxide as the raw material for sugar construction. This means they need to open their stomates to allow CO_2 to enter from the atmosphere. However, opening the stomates to dry desert air invites massive water loss. Some groups of plants evolved solutions to this problem by modifying photosynthesis:

- **C4 photosynthesis:** Plants that do C4 photosynthesis rely on an enzyme with high affinity for CO_2 to capture and concentrate CO_2 within their leaves by attaching it to a three-carbon compound called phosphoenolpyruvate (PEP), producing a four-carbon acid. They can move these acids to the cells doing photosynthesis, providing a source of carbon even when their stomates are closed to prevent water loss.
- **Crassulacean acid metabolism (CAM):** Plants that do CAM photosynthesis only open their stomates at night, when the risk of water loss is less. They also attach CO_2 to PEP, producing a four-carbon compound that they store in their vacuole. During the day, when sunlight is available, the four-carbon compounds break down, releasing CO_2 for photosynthesis within the plant cell.

Desert animals

In environments with plenty of water, animals can drink all they need and flush out any excess salts in their urine. Desert animals can't afford the luxury of producing large amounts of urine because they need to conserve as much water as possible. The kangaroo rat has solved this problem by evolving kidneys that are extremely efficient at concentrating salts. Human kidneys produce urine with solute concentrations about four times that of the blood. In contrast, the kidneys of kangaroo rats produce urine with fourteen times the solute concentration of their blood, enabling the rats to release their wastes with minimal water loss. Kangaroo rats are so good at water conservation that they can survive without drinking water and getting the moisture they need from their food.

Regulating body temperature

Life exists in every environment on Earth, from cold mountain snows to hot thermal springs. However, every living thing has a range of temperatures at which it can survive, and a temperature *optimum* at which it functions best. Changes in temperature can slow growth or even result in death because they impact the metabolism and structure of living things:

- » **Increases in temperature:** As temperatures increase, the *kinetic energy* of molecules increases, meaning that molecules move faster. If these increases are within the range of tolerance of an organism, they can speed up metabolism and even be beneficial. For example, low-grade fevers in humans are thought to speed up the activity of white blood cells needed to fight an infection. However, if the kinetic energy of molecules becomes too high, enzymes may lose their structure, or *denature*. This stops metabolism entirely, resulting in death.
- » **Decreases in temperature:** As temperature decreases, the kinetic energy of molecules decreases and metabolism slows. Organisms grow more slowly, which is why refrigeration helps protect your food from food spoilage organisms. If temperatures decrease below freezing, organisms face damage from ice crystals. Ice crystals are sharp and can literally tear cells apart, resulting in cell and organismal death. Organisms that survive exposure to freezing temperatures have adaptations to protect their cells. Some stock the cytoplasm of their cells with small molecules that act as antifreeze by lowering the concentration of water and preventing the formation of ice.

Some organisms have internal mechanisms to regulate body temperature; others are more affected by the environment:

- » Some organisms, including most mammals, are *homeotherms*, which means they can maintain their body temperatures within a certain range of environmental temperatures. You probably know that human body temperatures are typically around 98.6 degrees Fahrenheit (37 degrees Celsius). When the environment gets hot, bodies sweat to cool down. When it gets cold, bodies shiver, increasing kinetic energy and raising body temperature.
- » Some organisms, including many insects, frogs, and fish, are *poikilotherms* whose bodies change temperature to match their environment. When temperatures are cold, their metabolism slows, and they may be unable to move until the outside temperature increases.

Organisms that lack internal temperature regulation may have behavioral adaptations that compensate for environmental changes. *Ectotherms* use their environments to adjust their body temperature. When the air is cool, reptiles might seek out a sunny rock to lie on and absorb some heat. In the heat of the day, they might minimize sun exposure by seeking shade.

Endotherms modify their temperatures through their own internal metabolism. All energy transfers in cells produce heat, so heat production is a byproduct of metabolism. If you go outside on a day that is 70 degrees Fahrenheit, your body will still be around 98.6 degrees because of your metabolism. And if you start to exercise, increasing your metabolic rate, your body will start to sweat to cool you down from the excess heat production.

For you and other organisms, your ability to run your metabolism ultimately depends on your ability to find and consume food. In conditions when food isn't available, some endothermic animals reduce their metabolism to conserve energy.

REMEMBER

Torpor is a reversible state of inactivity and lowered body temperature.

Some animals have daily periods of torpor, like hummingbirds, which enter torpor at night. Others go into seasonal torpor, like bats in the winter. Both of these animals would starve to death if they couldn't hunt for food because their bodies would burn through their reserves trying to keep warm. A fungal infection, called white nose syndrome, is killing bats because it causes them to rouse from torpor. When the bats awaken, their bodies warm, burning through their fuel reserves and ultimately leading to starvation.

2

Looking at the Living World

IN THIS PART . . .

Learn how reproductive success is maximized

Discover how the planet is populated

Find your niche

IN THIS CHAPTER

» Sketching a life history

» Balancing survival and reproduction

» Exploring the role of the environment

Chapter 5 Figuring Out Life Histories

Evolution by natural selection favors the individuals who are best able to survive and reproduce in an environment. Many abilities can contribute to an organism's success, including the ability to find food, to flee from a predator, or to find a mate. These types of traits determine which individuals will reproduce. It might seem like the best strategy is to quickly have as many babies as possible. All you have to do, however, is look around to see that this isn't true. Some organisms, like the fruit flies in your kitchen, seem to follow this pattern, but others, like whales and elephants, typically only have one calf at a time. The diversity of reproductive strategies in the world suggests that more than one strategy can lead to evolutionary success. In this chapter, you'll consider how different components of life history must be balanced to achieve maximum reproductive success.

Components of Life Histories

Organisms live their lives in so many different ways. Fruit flies hatch, spend a week as maggots, then completely transform their bodies to become adult flies. Adult flies mature sexually at two days old, and females can lay up to 20 eggs at a time, producing up to 400 offspring in a lifetime. In contrast, orca whale females become sexually mature when they are 10–13 years old and typically have just one calf at a time. They're pregnant for 15–18 months and nurse their calf for at least a year. Scientists aren't sure about their lifetime reproduction, but estimate that they can have a calf a year every five years for a total of five calves. Evolution has influenced the life histories of both fruit flies and orcas, but the resulting reproductive strategies are completely different.

REMEMBER

A *life history* describes important landmarks in an individual's life, including age at maturity, number of offspring, and life span.

Establishing key points

Every life history has many components, but there are four that are especially important in terms of reproductive success. These four factors are illustrated in Figure 5-1:

- ***Maturity* is the average age when individuals in a species can first reproduce.** Organisms that reproduce early may ensure that they produce at least some offspring before they die, but those that delay may have time to accumulate more resources before having offspring. Some bacteria can reproduce in as little as ten minutes, fruit flies mature in two days, and African killifish can mature in as little as two weeks. At the other end of the spectrum are species like humans, which mature around 14 years old, elephants at 16, and Greenland sharks at 150!
- ***Parity* is the number of times an organism can reproduce in its lifetime.** Organisms typically follow one of two patterns:
  - ***Semelparity* is one big burst of reproduction in a lifetime.** It's sometimes nicknamed *suicide reproduction* because the parent often dies afterward. Pacific salmon will spend several years growing and maturing, then swim upriver to the place where they originally spawned. Once there, the female digs a nest in the rocks and releases thousands of eggs. At the same time, a male releases sperm to fertilize the eggs. Both adult fish die within a few days. As they decompose, they supply nutrients to the environment that will ultimately improve conditions for their young. Many insects are semelparous, producing juvenile forms that specialize in feeding before transforming into an adult form that's built to reproduce and then die. Likewise, annual plants reproduce and die at the end of their growing season.
  - ***Iteroparity* is multiple reproductive events in a lifetime.** Most vertebrates, including humans, are iteroparous, as are many perennial plants.

Species that reproduce more than once typically produce fewer offspring per reproductive event than do closely related species that are semelparous. It seems that these species save some resources for future reproduction rather than spending them all at once.

- ***Fecundity* is the number of offspring an organism typically has when it reproduces.** Often, species either produce many small offspring or a few larger offspring. Reproduction requires energy, and parents only have a certain amount of energy to give.
- ***Longevity* is the typical lifespan of an organism.** Some species only reproduce once, but for others, the longer they live, the greater the opportunity for reproduction. Many species reproduce throughout their lifetimes. Only a handful of animals, including humans, live well beyond their reproductive years.

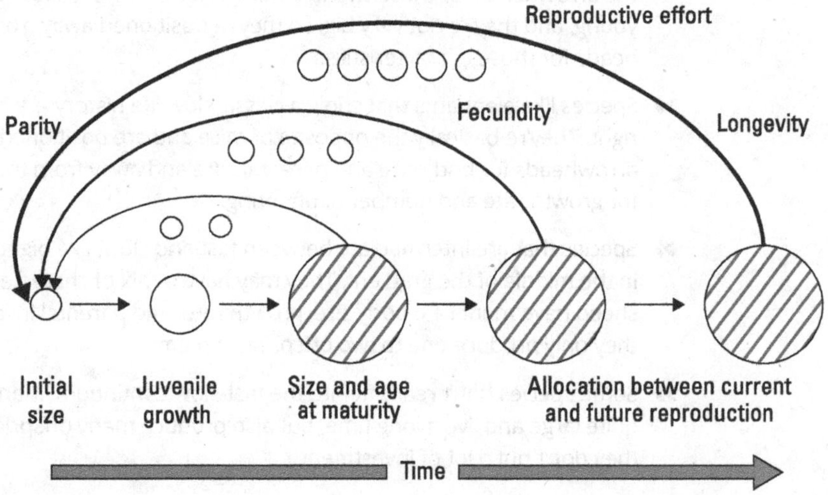

FIGURE 5-1: Key factors in a life history.

Life in the fast or slow lane

As scientists studied the life histories of different species, they found that many of them could be sorted into one of two groups:

- *Fast life history* species tend to have small bodies, shorter lifespans, and many offspring per reproductive event. Scientists also refer to these species as *r-selected species*.

- *Slow life history* species tend to be larger, live longer, and produce fewer offspring at a time. Scientists also refer to these species as *K-selected species.*

Although some species don't fit perfectly into these categories, many can be arranged along a *fast-slow continuum* as shown in Figure 5-2. The diagonal line running from upper left to lower right represents the continuum, or range, of life histories. The arrows around the outside of the figure indicate key life history characteristics. The direction of the arrows shows the value of each characteristic from least (no arrowhead) to greatest (arrowhead). Species are arranged spatially along these gradients:

- Species like mice or dandelions that have a classic fast life history are in the upper left. They have fast growth rates and lots of offspring, so they're near the arrowheads for those two factors. They don't invest a lot of care into their young, and they're not very big, so they're positioned away from the arrowheads for those characteristics.
- Species like elephants that show a classic slow life history are in the lower right. They're basically the opposite of mice and are positioned near the arrowheads for body size and parental care and away from the arrowheads for growth rate and number of offspring.
- Species that are intermediate between fast and slow, like bighorn sheep, are in the middle of the gradient. They may have a mix of characteristics. Bighorn sheep have moderate body size, growth rate, and parental investment, but they only produce one to two offspring at a time.
- Some species don't really fit into the fast-slow continuum. Many trees grow quite large and live a long time, but also produce many offspring, into which they don't put a lot of investment.

Thinking about species in terms of the fast-slow continuum provides a framework for thinking about how evolution shapes life histories and can provide a basis for predicting future events. For example, scientists found that the populations of fast species show greater fluctuation in population size in response to environmental change than do slow species. On the other hand, slow species tend to have higher rates of extinction than fast species. Being able to place a species along the fast-slow continuum can help identify which species are likely to be most vulnerable in the face of climate change (for more details on climate change, head to Chapter 12). It can also be used to identify useful species for restoration efforts (discussed more in Chapter 14).

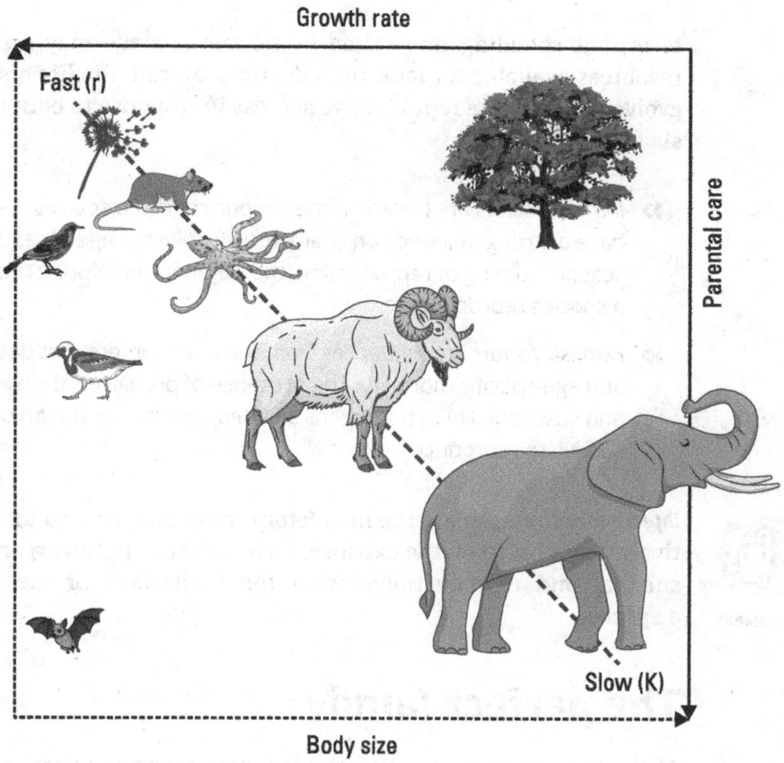

FIGURE 5-2: A comparison of fast and slow life histories.

Maximizing Lifetime Reproductive Success

Evolution by natural selection favors organisms with high reproductive success. In other words, those individuals that produce many offspring capable of survival and reproduction. If making babies is the name of the game, then why hasn't evolution produced a bunch of long-lived species that mature at just a few months old and produce tons of offspring at frequent intervals? Any working parent could answer this question: there's only so much energy to go around, so you constantly have to make decisions about how you spend it. Just as you might make trade-offs when you decide whether to go out to eat or pay the phone bill, life histories represent the trade-offs that occurred during the evolution of a species.

REMEMBER

Species have to make *trade-offs* because resources like time, energy, and materials used for one structure, function, or behavior can't be used for another.

The reason why all the species on Earth aren't just constant baby factories is that drawbacks balance the benefits of each life history strategy. For example, reaching sexual maturity quickly may be an advantage for reproduction in the short term, but spending more time in a juvenile stage may result in having more resources available for later reproduction. Overall, the life history of each species evolved to optimize reproductive success in response to both intrinsic and extrinsic factors.

- *Intrinsic factors* are based on the genetic code of a species. Genetic factors have a strong influence on characteristics like the age of sexual maturing, the seasonal timing of reproduction, the number of offspring produced, and how a species reproduces.
- *Extrinsic factors* are influences from the environment that determine survival and age-specific mortality. The presence of predators, the availability of food, and susceptibility to pathogens all affect whether or not an organism will survive to reproduce.

REMEMBER

Life history strategies are the life history traits that evolved to maximize reproductive success based on the extrinsic factors that affect an organism's probability of survival and reproduction within the limitations of the intrinsic factors of a species.

The perfect family

At the two extremes of the fast-slow continuum are species that produce many smaller offspring versus those that produce just a few larger offspring. This reflects the reality that sexual reproduction requires a lot of energy, especially for females whose bodies may support developing young by supplying food either directly or as stored food in eggs or seeds. In some species, parents use additional energy to care for their young for a period of time.

Species have evolved life history strategies to balance the benefits and costs of producing successful offspring. Although specific life histories are very different from each other, several overall patterns are consistent:

- **Species that invest more energy in their offspring produce fewer offspring overall.** This can be achieved either by raising just a few offspring at a time or by having longer periods between reproductions. Humans fall into this category, as do most larger mammals. In plants, species that make big seeds produce fewer seeds than those that make small seeds. The number of eggs laid by different bird species seems to match the number of young they can successfully support (see the sidebar "Pushing Parents to the Limit" for details).
- **Increasing the amount of energy input into individual offspring increases their fitness.** An ecologist named Barry Sinervo tested this idea by

manipulating the eggs of the western fence lizard. These lizards don't provide parental care after the eggs hatch, so parental energy is represented by the food reserves in the yolk of the egg. Sinervo's experiment had three treatments:

- For the experimental groups, Sinervo used a needle to remove a portion of the yolk from the eggs.
- For one type of control group, he didn't do anything to the eggs.
- For another type of control, he inserted a needle into the eggs but didn't remove any yolk.

The lizards that hatched from the eggs with less yolk were smaller and couldn't sprint as fast as the control lizards, which would make them more likely to be caught by predators. The lizards from both control groups were the same, indicating that the needle insertion itself didn't affect the developing lizards.

REMEMBER

Evolution by natural selection favors total reproductive success over a lifetime. The variation in life history strategies among species demonstrates that there's more than one way to maximize reproductive success. Successful strategies may involve producing many offspring all at once or producing fewer offspring multiple times over a long lifetime.

PUSHING PARENTS TO THE LIMIT

In the 1940s, ecologist David Lack suggested that birds lay the number of eggs that represents the best trade-off between the energy cost of reproduction and the benefit of successful offspring. He noted that birds that lived in regions with more daylight, and thus had more time to forage, tended to produce more eggs per clutch than birds that lived in areas with shorter daylengths. This suggested that the number of eggs in a clutch was limited by the amount of energy the parents had access to in the form of food.

Many ecologists have done experiments to test Lack's hypothesis. One simple way to test whether birds limit the number of eggs they lay is to remove eggs from the nest. In response to egg removal, some species of birds will continue to lay eggs, confirming the fact that birds don't just lay as many eggs as they physically can. However, in experiments with gulls, terns, and zebra finches, ecologists showed that eggs produced in response to removal experiments declined in nutritional quality. In gulls, the chicks that hatched from later eggs were less likely to survive to fledging than those from earlier

(continued)

(continued)

eggs. This supports Lack's idea that clutch size reflects adaptation for maximum success. In other words, birds may lay the largest clutch they can that will result in healthy offspring that are likely to survive.

Some experiments that tested Lack's hypothesis also included measurements of the impact of clutch size on parents. In experiments with small hawks called European kestrels, ecologists added eggs to the nests of some kestrels, removed eggs from others, and left some untouched. The amount of time that kestrel parents spent hunting for food correlated with the number of young in the nest. In other words, kestrel parents with more offspring spent more time in high-energy activities of flying and searching for prey than did parents with fewer offspring. Female parents who raised enlarged broods lost more weight during the breeding season than did females that raised normal or reduced broods.

Also, when ecologists evaluated tagged adult kestrels during the next breeding season, they found that the survival of birds that raised the enlarged broods was significantly lower than that of kestrels that raised normal or reduced broods. These results show that the demands of parenting required significant energy inputs from adult birds and negatively impacted their survival. The offspring from the enlarged broods didn't do well either: they grew more slowly and had a greater chance of dying than the fledglings from the normal and reduced broods.

Overall, these experiments support the idea that clutch sizes in birds are smaller than what is physically possible because they represent the result of evolutionary optimization for the total reproductive output of successful offspring.

When are you having kids?

The timing of reproduction for maximum success also involves balancing positive and negative factors. Organisms that reproduce early may have short-term success but may face consequences that limit their potential for long-term reproductive success:

- **Investment in offspring can lower the survival rate of parents.** An organism that reproduces early but doesn't survive very long may not produce as many offspring overall as compared to a long-lived species that delays reproduction. In experiments with small hawks called European kestrels, ecologists showed that increased parental investment in their chicks resulted in lower parental survival (for more details, check out the "Pushing Parents to the Limit" sidebar).

» **Organisms that grow for a longer period before reproducing may be able to produce more offspring.**

- Increasing body size generally enables an organism to produce more offspring at one time.
- The energy cost of one reproductive event can negatively impact subsequent events.
- So, for organisms with longer lifespans, the best strategy for lifetime reproductive success may be to delay reproduction.

REMEMBER

Species with shorter lifespans tend to reproduce earlier in life than those with longer lifespans.

Growing older

The evolution of lifespan may involve a trade-off between the benefits of increased reproductive lifespan and the cost of increased age-dependent mortality. Most organisms experience a decline in fecundity as their physiological functions begin to deteriorate with age.

REMEMBER

Senescence is the age-related decline in an organism's ability to survive and reproduce.

Natural selection favors traits that improve biological fitness, but there is no obvious adaptive advantage to senescence. This has led to many different proposed explanations for why it exists. Here are just a few:

» **Wear and tear:** DNA mutations are a fact of life. The longer an organism lives, the more mutations it will accumulate in its DNA. If an organism lives long enough, the effect of those mutations begins to show.

» **Trade-off:** Senescence is part of a life history strategy that was advantageous overall. Investing resources in reproduction may be balanced by letting the body decline after reproduction is complete. Or perhaps an adaptation that improved fitness in a particular environment may have been somehow linked to senescence.

» **Disposable soma:** This phrase basically means that the parts of your body that aren't needed for reproduction are not worth protecting. Natural selection favored a strategy that focuses most of your energy on protecting the DNA in your reproductive cells from damage, but doesn't worry too much about what happens to the rest of your cells. Mutations accumulate, and you're back to wear and tear.

» **Sex sells:** Your traits when you are sexually mature can determine your reproductive success, so natural selection has a large impact on shaping those traits. After you're done reproducing, natural selection has little impact.

The timing and rate at which senescence occurs aren't equal across species, which raises more questions about how natural selection has shaped this process. Some organisms even seem to be resistant to senescence, including perennial plants that show very few signs of aging, and a jellyfish nicknamed The Immortal Jellyfish because it can reverse the aging process (for more information, check out the sidebar "Do You Want to Live Forever?").

REMEMBER

The mechanisms that repair damage and thus slow senescence are under genetic control and can be modified by evolution.

DO YOU WANT TO LIVE FOREVER?

Can you imagine what it would be like if you could suddenly decide that you didn't want to age anymore, so you dissolved into a puddle of goo and then started life over as a baby again? If your name were *Turritopsis dohrnii*, aka the immortal Jellyfish, you could do just that.

The bell-shaped swimming creature that you probably imagine when you hear jellyfish is the adult form of the animal, called the *medusa*. Jellyfish start life as small flat larvae that land on a surface and grow into a stalked form with tentacles called a *polyp*. The polyps can grow for a long time and reproduce asexually. When conditions are right, the polyps produce medusae for sexual reproduction. Medusae channel most of their energy into reproduction and don't usually live very long after they release their sperm or eggs. Thus, the jellyfish stage is the end of life for this animal.

You might be able to imagine the surprise of two researchers, Giorgio Bavestrello and Christian Sommer, when they returned to the lab after the weekend and found that the jellyfish they'd been studying had turned back into polyps. If the jellyfish had reproduced normally, they might have expected to see larvae, but no one had ever seen an animal switch from a sexually mature stage to a juvenile stage. The scientists continued to watch their cultures and discovered something completely new. The medusae were turning into a cyst form that settled to the bottom of the tank and then grew into a polyp. The jellyfish seemed to be reversing the aging process.

By studying these jellyfish, scientists are gaining insights into aging and development. The cells in the cysts silence many genes associated with cell differentiation, the process that causes cells to specialize for different functions. It's almost as if the cells are becoming like human stem cells, capable of becoming any type of cell. The cells in the cysts also increased the activity of genes that repair DNA. Likewise, cells from *T. dohrnii* have more copies of genes that repair and protect DNA than do cells from a closely related jellyfish that doesn't reverse development. These observations suggest that protecting and repairing DNA may be key to the process of biological immortality. Before you worry about the potential for these jellyfish to take over the world, or at least the oceans, the scientists pointed out that these jellyfish can still die from predation and disease.

Exploring the Impact of the Environment

Evolution by natural selection influences the genetic foundation that underlies life history strategies. In addition to genetic constraints, life histories can also be influenced by the environment. Changes in environmental factors like temperature, food availability, and predation result in both long-term and short-term responses in life histories:

- Over the long-term, evolution may lead to changes in the genes that regulate the life history of a species.
- In the short-term, species may adapt by:
  - changing feeding behavior or growth rates.
  - storing resources during favorable environmental conditions to survive less favorable times.
  - adjusting the timing of life history stages.

REMEMBER

Genetics and the environment both influence life history strategies.

You can only reproduce if you're alive

Life history strategies evolve to maximize lifetime reproduction success. The strategies of fast species rely on the production of many offspring as quickly as possible, whereas slow species maximize reproduction over longer periods of time. Ecologists hypothesized that fast life histories might be more likely to evolve in circumstances where individuals were under higher threat, for example, by predators. If individuals in a population could die at any time, then those individuals that reproduce more quickly would be more likely to survive.

To test this idea, ecologist Stephen C. Stearns exposed populations of fruit flies to two different conditions:

- » Some fruit flies were raised in high mortality conditions, where 90 percent of the population was killed twice per week.
- » Other fruit flies were raised in low mortality conditions where only 10 percent of the population was killed twice per week.

After four years of ruthlessly exterminating fruit flies, Stearns compared the fruit fly populations and found that the life histories of the two groups, which were the same at the start of the experiment, had diverged into two different strategies. The fruit flies from the high mortality conditions reached maturity faster, laid larger numbers of eggs earlier, and had shorter lifespans than the flies from the low mortality conditions.

Similarly, ecologist David N. Reznick showed that guppies that live with high predation matured earlier, put more resources into reproduction, and produced greater numbers of smaller offspring than did guppies that live with lower predation.

REMEMBER

These experiments prove that evolution shapes life history strategies in response to environmental conditions.

Finding food

Organisms must have food to survive and reproduce. Species that must forage for food use behavioral strategies to maximize gain and minimize risk.

REMEMBER

Optimal foraging refers to the strategy employed by organisms to maximize energy gains while minimizing energy costs during foraging.

One factor that can affect the energy balance of foraging is the distance that an animal needs to travel to obtain food. The farther an animal travels, the more area they can search, and the greater the possibility of high rewards. On the other hand, greater travel means a larger energy cost and a potential reduction in net energy gain.

Another factor that influences foraging behavior is the risk of becoming prey while out hunting for food. Animals may avoid areas that have high-quality forage because they are too exposed to potential predators. In Poland, deer grazing patterns were different in tree plantations with and without wolves, suggesting that fear of predation influenced foraging behavior.

Feast and famine

Many species store resources when things are good so they can survive the hard times. You're probably familiar with squirrels hiding nuts in the Fall or bears putting on weight so they can survive the winter. These benefits of these behaviors seem obvious, but what about the cost? An animal that saves food for later doesn't get the benefit of those calories when the food is harvested, which means there is a trade-off between immediate gain and future benefit. Likewise, animals that store fat may become slower and less able to escape predators, taking a risk in the present that may pay off in the future.

Another strategy for surviving periods of time when food or water is scarce is for species to go dormant.

REMEMBER

Dormancy is a period of physiological inactivity and reduced metabolism.

Many species that live in environments with seasonal weather changes show some form of dormancy. Ground squirrels, bears, and marmots all *hibernate*, going into a state of deep rest that lowers their heart rate, breathing, and metabolism. Bats go through shorter periods of lowered metabolic rate called *torpor*. Some insects enter *diapause* before the start of winter, producing a sort of antifreeze to keep water in their bodies from turning to ice and dropping their metabolic rate. Perennial plants may also produce chemicals to protect themselves from freezing and reduce their metabolic rate before winter. When organisms are dormant, they can't acquire food to support their reproductive efforts, so dormancy only makes evolutionary sense in areas that experience periods of food shortage.

IN THIS CHAPTER

» Describing population structure

» Modeling population growth

» Looking at population distribution

Chapter 6

Painting a Picture of Populations

Individuals reproduce to form groups of the same species called *populations*. As shown in Figure 6-1, populations of one species combine with those of other species to form *communities* (Chapter 8). Communities interacting with each other and their environment form the *ecosystems* that support all life on Earth, which is collectively called the *biosphere*. Understanding the factors that influence populations helps ecologists predict how they might change in the future. In particular, understanding the size and distribution of populations across a range of potential habitats is important for the design of strategies to conserve threatened species.

REMEMBER

Population ecology is the study of how and why populations change over time.

In this chapter, you'll explore the factors that contribute to the growth and decline of populations.

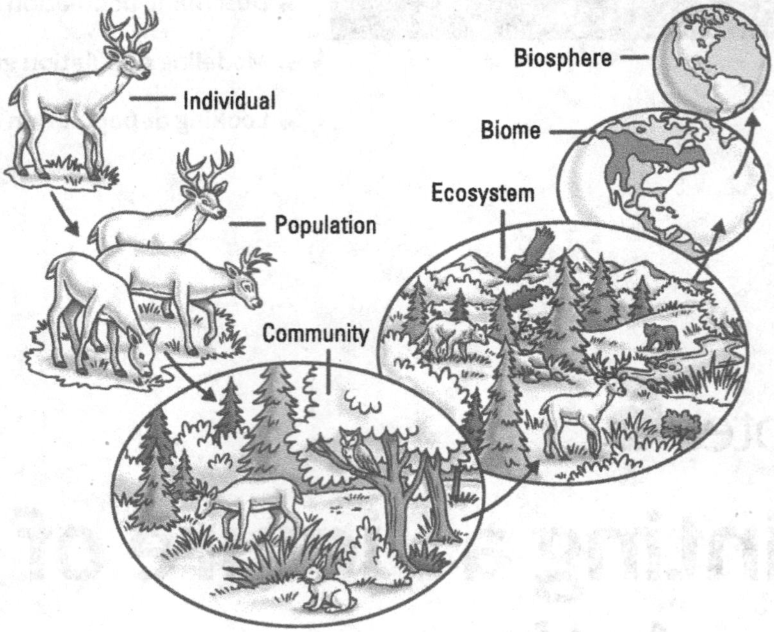

FIGURE 6-1: Levels of complexity of living organisms on Earth.

Growing a Population

Populations are important to the biological fitness of individuals because they represent a group of potential mates. When individuals have access to a large pool of potential mates, it's more likely that their offspring will be genetically diverse, increasing the chances that some individuals will survive in the face of environmental change like global warming (see Chapter 12 for details). Ecologists study the fundamental characteristics of populations to understand the factors that maintain population resiliency.

REMEMBER

Population structure describes key features of populations includes population size, density, distribution of individuals, the ratio of males to females, and the age structure of the population.

Measuring abundance

Population size is the number of individuals in the population, usually represented by the symbol *N*.

Two populations of the same size might seem very different if they occupy different areas. For example, New York City has a population of around 8.5 million people. The entire state of Virginia also has a population close to 8.5 million. If you visited New York City and Virginia, you'd probably feel like New York City was much more crowded. That's because the 8.5 million people in New York City occupy a much smaller space than that of Virginia. In other words, the population in New York City is much more dense.

REMEMBER

Population Density is the number of individuals in the population divided by the area the population occupies.

The population of humans in New York City is 8,478,072 people. Divided by 300.46 square miles, that's a population density of 28,217 people per square mile. In contrast, the human population of Virginia is 8,631,393. Divided by 39,490 square miles, that's a population density of about 219 people per square mile. Two populations that aren't that different in size seem very different when you compare their densities.

Monitoring growth

Populations constantly change as new individuals are born and others die, or as individuals migrate (see Figure 6-2):

- Birth, or *natality*, adds individuals through reproduction.
- *Immigration* adds individuals when they migrate in from other populations.
- Death, or *mortality*, removes populations by predation, disease, or senescence. (See Chapter 5 for more details on senescence.)
- *Emigration* removes individuals when they migrate out to other populations.

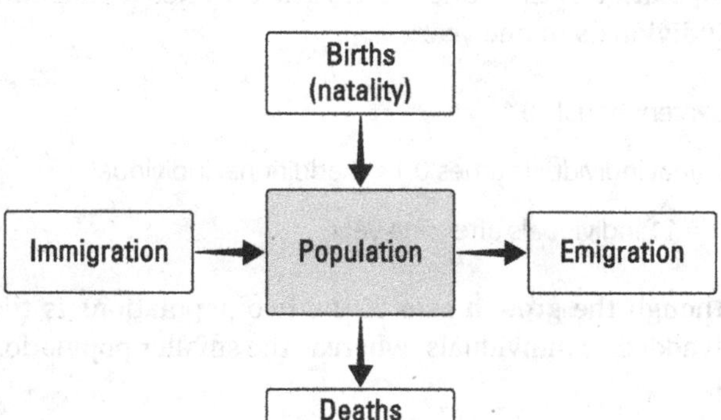

FIGURE 6-2: Change in population size.

REMEMBER

Population dynamics studies how and why populations change in size and structure over time.

TIP

You can remember the difference between immigration and emigration by thinking that *i* (immigration) brings individuals *i*nto the population, and *e* (emigration) helps individuals *e*xit.

Ecologists use mathematical equations and computer models to track populations and predict how they will change. The change (Δ) in population size (N) over a given period of time (*t*) is represented as:

$$N_{t+1} = N_t + B_t - D_t + I_t - E_t$$

In other words, population size at a future time interval (t+1) is equal to the original population size (N_t) plus the births (B_t) that occurred during that time, minus the deaths (D_t), plus the immigration (I_t), and minus the emigration (E_t).

Population growth rate (r) is the average change in the population over time. It's expressed on a *per capita* basis; in other words, as the rate of growth per individual in the population. Larger populations typically increase in number faster than smaller populations because they have more individuals capable of reproducing. For example, assume you have two populations of the same species that have a per capita growth rate of 10 percent per year. If one population consists of 100 individuals, then in one year, the population would increase to 110 individuals:

1. Ten percent equals 0.1.
2. 100 original individuals times 0.1 = 10 additional individuals.
3. 100 + 10 = 110 after one year.

Another population of the same species that only has 10 individuals would increase to just 11 individuals in one year:

1. Ten percent equals 0.1.
2. 10 original individuals times 0.1 = 1 additional individual.
3. 10 + 1 = 11 individuals after one year.

So, even though the growth rate of the two populations is the same, the larger population added 10 individuals, whereas the smaller population only grew by one individual.

TIP

The per capita growth rate has the same effect as the interest rate on a bank account: if you have a lot of money in the bank, the amount you gain by interest will be greater than if you have a small amount of money in the bank.

Looking at life tables

Ecologists collect information about birth and death for different species and organize them into *life tables*. These tables are helpful for the science of *demography*, or the statistical study of populations over time, including birth rates, death rates, and life expectancies.

TIP

Insurance companies use life tables for human populations to calculate the risk of injury, illness, and death so they can set their prices accordingly.

One type of life table is a *cohort life table*, which tracks a group of individuals born at the same time (the cohort) over a period of time. An example of a cohort life table for song sparrows is shown in Table 6-1.

TABLE 6-1 A Cohort Life Table of Song Sparrows

Age in Years (x)	Number Alive (n_x)	Survivorship (l_x)	Number Dying per Interval (d_x)	Mortality Rate (q_x)
0	115	1	90	0.78
1	25	0.217	6	0.24
2	19	0.165	7	0.37
3	12	0.104	10	0.83
4	2	0.017	1	0.50
5	1	0.009	1	1
6	0	0		

The number of the population that is alive at each age is counted by ecologists using methods like mark and recapture or random sampling (see Chapter 15 for details). From the number of the cohort that's alive each year, other demographic information can be calculated:

- **Survivorship (l_x) is the proportion of the original cohort that survives to age (x).** It's calculated by dividing n_x by n_0. In other words, $l_x = n_x / n_0$.
- **The number dying during a time interval (d_x) is the number of individuals that died between time x and x+1.** It's calculated by subtracting n_{x+1} from n_x. In other words, $dx = n_x - n_{x+1}$.

» **The mortality rate (q_x) is the proportion of individuals who died during a time interval.** It's calculated by dividing the number of individuals that died in an interval (d_x) by the number of individuals at the beginning of that interval (n_x). In other words, $q_x = d_x / n_x$.

Life tables provide useful insights into the life histories of populations. They help identify windows of high mortality and high reproductive rates. Researchers may follow up on these insights to investigate the causes of mortality. Information on mortality and fecundity can also be used to develop plans to protect endangered species.

Data from life tables can also be used to compare survivorship in different populations, as shown in Figure 6-3. In general, species fall into one of three different patterns of survivorship:

» **Type I survivorship:** Offspring and juveniles have high survival rates so that organisms live out most of their life span, dying in old age. Humans have a type I survivorship because most individuals survive to middle age and beyond.

» **Type II survivorship:** Death occurs randomly throughout the life span, usually due to predation or disease. Some birds show Type II survivorship because they are constantly at risk from cats and other predators.

» **Type III survivorship:** Offspring and juveniles have high mortality rates, and few members of the population survive to reproductive age. However, individuals that do survive to reproductive age often live out the rest of their life span and die in old age. In other words, Type III organisms die young. Most insects show Type III survivorship because they go through several developmental stages of larvae and instars before they become adults that can reproduce. Birds and spiders prey heavily on the early stages, so that only a small percentage make it to adulthood..

Species do not always fit perfectly into these three types of survivorship. For example, two populations of the same species, one that lives in an environment with predators and one that doesn't, might show different patterns. Observations on the differences between populations of the same species can lead ecologists to propose hypotheses and investigate the factors influencing mortality.

The data from life tables can be used to create a visual model of a population called an *age-structure diagram*, like the ones shown in Figure 6-4. These diagrams group individuals of the same age into *age classes*, then represent the number of individuals within each class by the width of the layer in the diagram. The diagrams separate males and females so that the numbers of each sex within a class can also be seen.

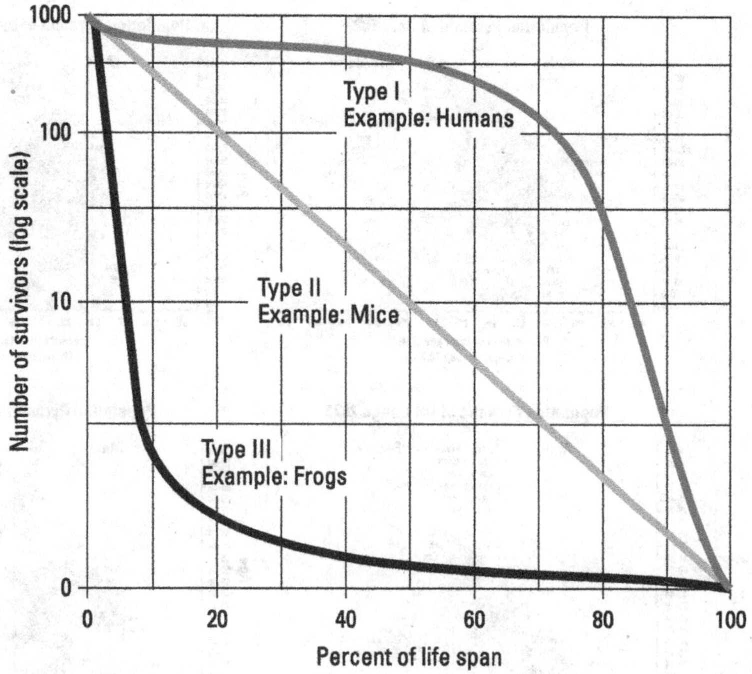

FIGURE 6-3: Life tables.

Age-structure diagrams, also called *population pyramids*, create a rapid snapshot of the population that reveals its age profile as well as its size. Many species reproduce at specific times in their life history, so an age-structure diagram can give you a sense of whether a population is likely to grow, stay the same, or shrink based on the number of individuals within the reproductive age classes. Figure 6-4 illustrates this by comparing three human populations from the year 2023:

- **A pyramid-shaped age-structure diagram indicates a rapidly growing population.** The age-structure diagram for the Democratic Republic of the Congo has a wide base and a narrow top because most people are below reproductive age. The younger age classes are larger than the reproductive age classes, so the population is growing.
- **An evenly shaped age-structure diagram indicates the population is relatively stable.** The number of people at and below reproductive age in the United Kingdom is about equal, with a decrease in the older age classes. Because the younger age classes are about the same size as reproductive age classes, the population is stable.
- **An age-structure diagram that has a smaller base than the middle portion indicates the population is decreasing in size.** Italy has more people in the age classes older than reproductive age. The youngest age classes are smaller than the reproductive age classes, so the population is decreasing.

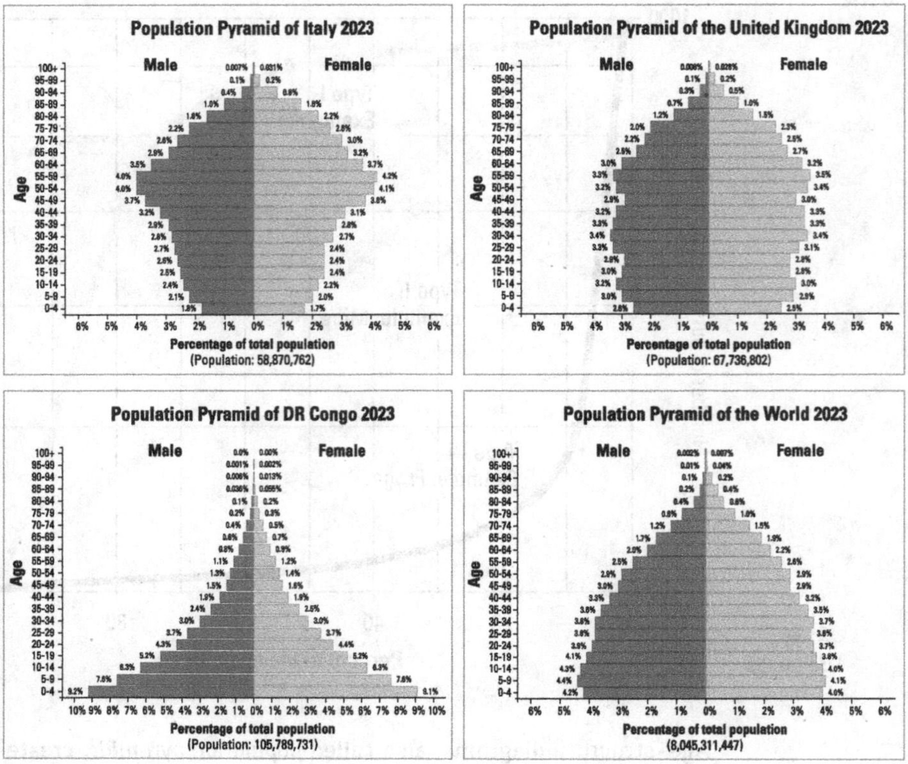

FIGURE 6-4: Age-structure diagrams break down age groups in populations.

Age-structure diagrams are particularly useful for visualizing human population growth around the world. The human population has passed 8 billion (see Figure 6-5) and is projected to keep growing for another 50-60 years if nothing changes. The United Nations predicts that the population will reach 10.3 billion by 2080. (For more details on these types of predictions, see the sidebar "How Many People Are Too Many?") The growing human population is consuming vast amounts of resources, putting stress on planetary systems (see Chapter 12), and threatening the survival of other species (see Chapter 13). The good news is that the age-structure pyramid for the global human population is changing from a triangular shape with most individuals in younger age classes to a more columnar shape with even numbers between young and reproductive age classes (see Figure 6-4). In other words, the global population is shifting away from rapid growth to a steady state. After 2080, the UN predicts that the global population will shrink a small amount, then remain steady.

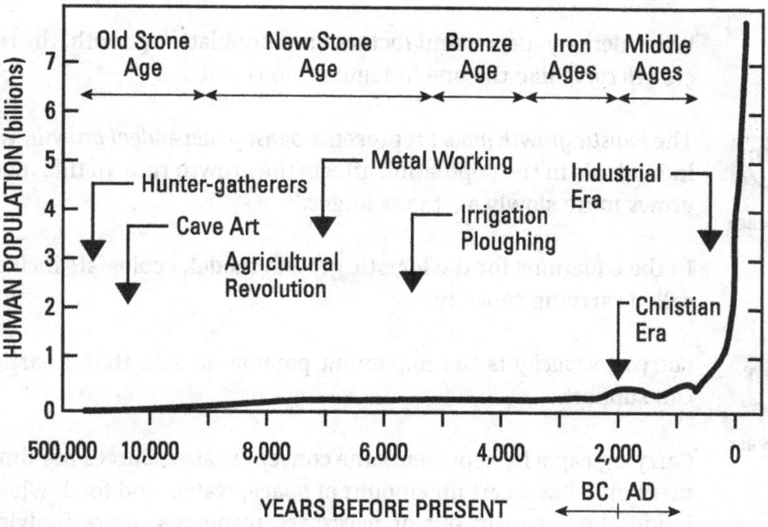

FIGURE 6-5: Human population growth.

Limiting growth

The *J-shaped growth curve*, like the one in Figure 6-6 (left), shows what unlimited growth of a population looks like when graphed as a function of time. The graph shows a *J-shaped curve* because the growth rate of the population accelerates. Ecologists call this type of growth *exponential growth* because the population grows as an exponential function of time.

Exponential growth rarely occurs in nature. Populations may show exponential growth for short periods of time, but then environmental factors typically act to limit their growth rate. Even bacteria, which are champions of exponential growth, slow down when they begin to run out of food.

Population growth can be limited by a number of environmental factors, which population ecologists group into two categories:

- **Density-dependent factors** are more likely to limit growth as population density increases. For example, large populations may not have enough food, water, or nest sites, causing fewer organisms to survive and reproduce. Crowded populations may also be more susceptible to disease and predation. Lower birth rates combined with higher death rates slow population growth.
- **Density-independent factors** limit growth but aren't affected by population density. Changes in weather patterns that cause droughts or natural disasters, such as earthquakes, kill individuals in populations regardless of that population's size.

When density-dependent factors limit population growth, the result is an *S-shaped growth curve* like the one in Figure 6-6 (right).

REMEMBER

The *logistic growth model* represents *density-dependent growth*, when the number of individuals in the population affects the growth rate. In this model, the population grows more slowly as it gets larger.

In the equations for the logistic growth model, ecologists include a limiting factor called carrying capacity.

REMEMBER

Carrying capacity is the maximum population size that a particular environment can support.

Carrying capacity represents the concept that resources are limited. Any environment only has a certain amount of space, water, and food. When a growing population starts to run out of necessary resources, more individuals will die, and fewer will be born, ultimately slowing the growth rate. The population size may stabilize when the carrying capacity is reached, as shown in Figure 6-6 (right).

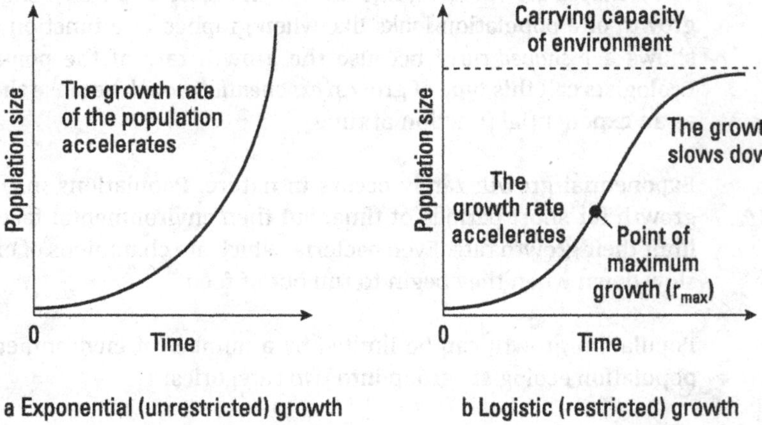

FIGURE 6-6: Exponential and logistic growth models.

In nature, population growth doesn't always fit the logistic growth model. Populations may grow above carrying capacity, then decline due to density-dependent factors, resulting in an uneven wave that hovers around carrying capacity. If the habitat becomes damaged, then the carrying capacity of an environment may be reduced, resulting in population decline.

Predicting human population growth is challenging because changes in technology, education, agriculture, and other factors can affect how different human populations grow. Many rich industrialized nations have reached a stage of *zero population growth*, where birth and death rates are equal in these countries. But, although birth rates are declining globally, some poorer, less industrialized nations have higher birth rates relative to their death rates, plus a lot of younger people, resulting in growing populations. Factors that lower birth rates include:

- Increased access to education, particularly for women.
- Economic factors, such as access to jobs and careers, and the cost of raising children.
- Changes in norms and values regarding family size and the role of women in society.

Because human populations around the planet are unique, population ecologists have developed a special *demographic transition model* to note the stages of development countries go through on their way to population stabilization (see Figure 6-7 for the visual):

- **Stage 1:** Birth and death rates are both high. Basic sanitation and modern medicine aren't yet available to lower the death rate and extend the life span. Demographic transition has not yet occurred.
- **Stage 2:** Sanitation and medicine lower the death rate, but the economy and social factors still encourage a high birth rate. In the Democratic Republic of the Congo, shown in Figure 6-3, large family size is celebrated, girls marry young, and contraception isn't readily available.
- **Stage 3:** Increased urbanization reduces the need for large families, and the cost of raising and educating children encourages fewer births. The birth rate drops and becomes close to that of the death rate. The population still grows, however, as earlier generations reach reproductive age. As countries pass from stage 2 to stage 3, they make a partial demographic transition.
- **Stage 4:** The population becomes stable, and birth rates equal death rates. When countries reach stage 4, they've made a full demographic transition.
- **Stage 5:** In some models, birth rates fall below death rates, and the population declines.

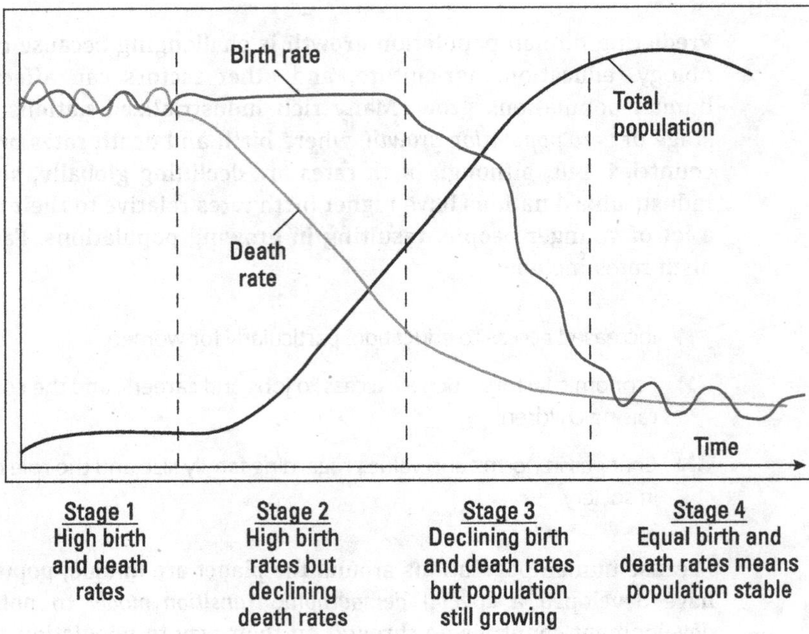

FIGURE 6-7: The demographic transition model.

HOW MANY PEOPLE ARE TOO MANY?

In 1920, two scientists named Raymond Pearl and Lowell J. Reed developed and tested a logistic growth model to see if it could explain and predict the growth of the human population of the United States of America. The calculations from their model matched very well with actual population census data for the years between 1790 and 1910. Because their model fit the actual data very well, the scientists used it to make some predictions. First, they inferred that the *inflection point* of the logistic growth curve, which represents the point of *maximum growth rate* (see Figure 6-6 [right]) had already occurred in 1914. They predicted that the census data for 1920 would show that the growth rate of the US population was already declining. Second, they estimated that the maximum population of the United States would be 197,274,000 people.

The population of the United States is currently 342,442,380 people, so it seems that the logistic growth model proposed by Pearl and Reed wasn't very accurate in its ability to predict into the distant future. It did predict US population growth very well until about 1950, which was 30 years after they first published their analysis. They were aware of the limitations of their model and had the incredible foresight to comment on the disparity between their predicted maximum population and the ability of the United States to sustain that population. Based on typical American calorie consumption and agricultural food production, they predicted that the US would have to import food unless people could be persuaded to change their food habits. They even predicted potential problems with obtaining food to import from other countries that would have to feed their own growing populations.

Since Pearl and Reed, many other scientists have tried to think about human population growth in logistic terms and predict the carrying capacity of the environment for human beings. Recent estimates by Dutch scientists for how many humans the Earth can support ranged from 500,000,000 to 1,000,000,000,000,000,000,000. (That's 1 *sextillion*!) Clearly, these scientists are including different variables in their models!

One of the reasons it's difficult to predict a specific carrying capacity for humans is that our relationship to our environment constantly changes. For example, the human ability to extract and use resources from the environment has increased many times, including increased food production during the Agricultural Revolution and increased ability to use fossil fuel energy in the Industrial Revolution (see Figure 6-6 for timing). Innovations like these have the potential to increase the carrying capacity of the Earth for humans.

Instead of trying to estimate carrying capacity, many ecologists and environmentalists have switched to thinking about the ecological footprint of human populations. Your *ecological footprint* is the total area of productive ecosystems required to support your lifestyle, typically reported in units of how many Earths would be needed if everyone lived like you. People who are vegetarians and ride bicycles have smaller ecological footprints, so one Earth might be enough to support 8 billion people living that way. At the other extreme, seven Earths might be required to support 8 billion people who eat beef, have large homes, drive energy-inefficient vehicles, and fly often. You can go to `https://www.footprintcalculator.org` to determine your ecological footprint.

Spreading Out into the Environment

Populations distribute themselves into environments that support their growth. The spread of a population may be limited by abiotic factors like climate and soil type, geographic barriers like rivers or mountains, or by biotic factors such as competition or predation.

REMEMBER

The *distribution* of a population is the geographical area occupied by that population. In other words, it's how far the population is spread out over the landscape. Ecologists also refer to distribution as the *geographic range* of the population.

Distribution

When thinking about the distribution of population, ecologists sometimes contrast the area that could support the population with the area actual occupied by the population:

- The *fundamental niche* is the range of physical conditions in which the population can persist.
- The *realized niche* is the actual environment occupied by the population. For migratory populations, the realized niche includes all the areas occupied by the species during their life history.

By comparing the components of the fundamental niche and realized niche, ecologists can identify factors that limit the distribution of a population, such as predators, diseases, or competitors.

Sometimes, populations can't access suitable habitat because of *dispersal limitations*. A species that lives in a biome in Asia might have the ability to live in the same biome in North America, but its progeny have no way to get there. As humans traveled around the world, however, they often took species with them, either on purpose or by accident. Coffee plants, which are native to Africa, now grow in Central and South America because humans transplanted them around the world. Unfortunately, some species become invasive when they're transplanted to new environments because they don't always encounter the same limitations to their growth (see Chapter 13 for more details on invasive species).

Dispersion

REMEMBER

Population ecologists use the term *dispersion* to describe the spacing of a population within their geographic range. Populations disperse in three main ways, which are illustrated in Figure 6-8:

- **Clumped dispersion:** In this type of dispersion, most organisms are clustered together with a few organisms in between. Examples include people in New York City, herds of elephants, and bees in a hive.
- **Uniform dispersion:** Uniformly dispersed organisms are spread evenly throughout an area. Grapevines in a vineyard and territorial birds like Emperor penguins are examples of uniform dispersion.
- **Random dispersion:** In this type of dispersion, one place in the area is as good as any other for finding the organism.

REMEMBER

Random dispersion is rare in nature but may result when seeds or larvae are scattered by wind or water. Examples of random dispersion include barnacles scattered on the surfaces of rocks and plants like dandelions with wind-blown seeds settling down on bare ground.

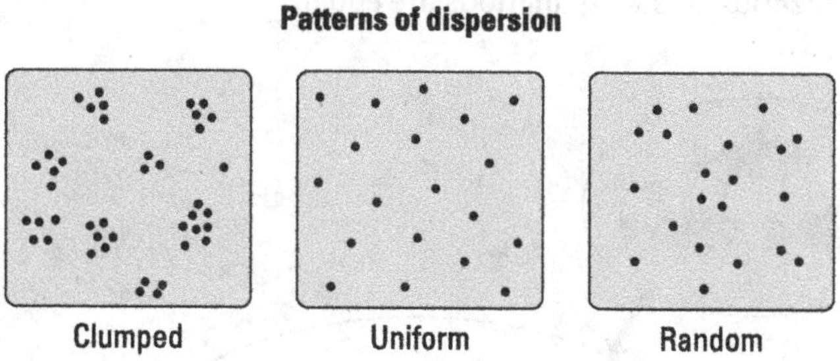

FIGURE 6-8: Population dispersion.

Metapopulations

Sometimes, populations are more of a network than a single unbroken entity. Subunits, called *subpopulations*, link together to form a larger metapopulation. Figure 6-9 shows a metapopulation of fish that includes four subpopulations.

REMEMBER

A *metapopulation* is a group of subpopulations that occupy spatially distinct habitat patches but can immigrate and emigrate between patches.

Metapopulations function as a single population because individuals from different subpopulations can interact and reproduce with each other, enabling gene flow within the metapopulation.

Metapopulations sometimes occur between islands within a chain, or between islands and the mainland. They may also result from human activities that disrupt ecosystems, including road building, dam construction, agriculture, and urban development.

REMEMBER

Habitat fragmentation occurs when a large habitat is transformed into a group of disconnected fragments.

Metapopulation structure can help stabilize a population because one subpopulation can supply new individuals to another. If one subpopulation goes extinct, the unoccupied patch may be recolonized from another subpopulation. In some cases, a large subpopulation with a positive growth rate may act as a *source* of individuals for smaller *sink* subpopulations that may be shrinking. For example, a larger mainland population may supply colonizers to nearby islands. If an island subpopulation is about to go extinct, the larger population may exert a *rescue effect* by supplying new individuals. These stabilizations are only possible if dispersal routes between subpopulations remain open.

REMEMBER

A *dynamically stable* metapopulation results when the rates of extinction and recolonization of subpopulations are equal.

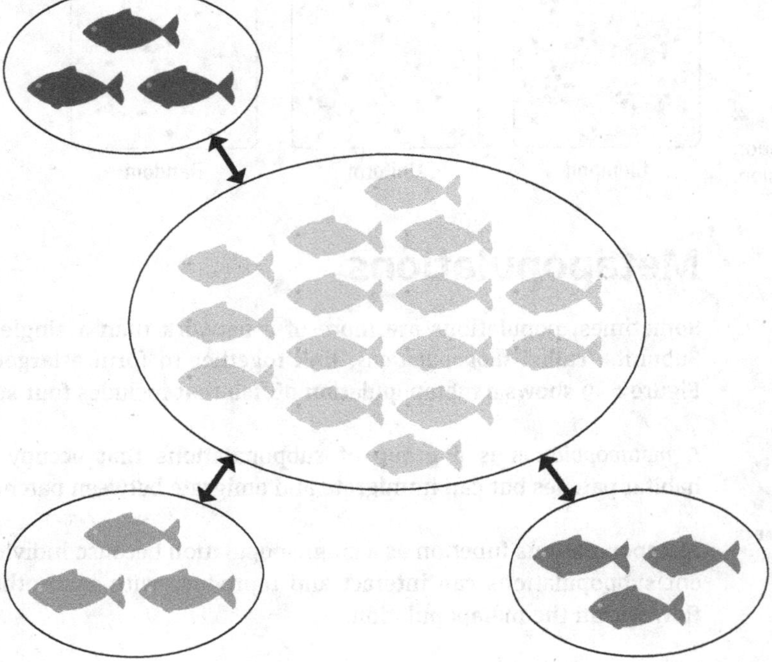

FIGURE 6-9: Metapopulations in fish.

Diving Deeper into Population Dynamics

Another way of thinking about population growth is to say that it is directly proportional to population size. Ecologists write this relationship with an equation:

$$dN/dt = rN$$

REMEMBER

The growth rate of the population (dN/dt) is directly proportional to the size of the population (N), multiplied by the per capita growth rate (r).

If the population grows without limits, the number of individuals added during each time period will keep increasing even though the per capita growth rate stays the same.

Graphing growth

In order to calculate the change in population size over time, ecologists used calculus to derive the following from the equation for population growth rate:

$$N_t = N_0 e^{rt}$$

This equation says that the population size at time t (N_t) is equal to the population size at time zero (N_0) multiplied by the constant e raised to the power of the per capita growth rate (r) multiplied by time (t). Euler's number (e) is a mathematical constant with a value of about 2.72. You can find e on a scientific calculator.

Figure 6-10 shows what unlimited growth of a population looks like when graphed as a function of time. The graph shows a *J-shaped curve* because the growth rate of the population accelerates. Ecologists call this type of growth *exponential growth* because the population grows as an exponential function of time.

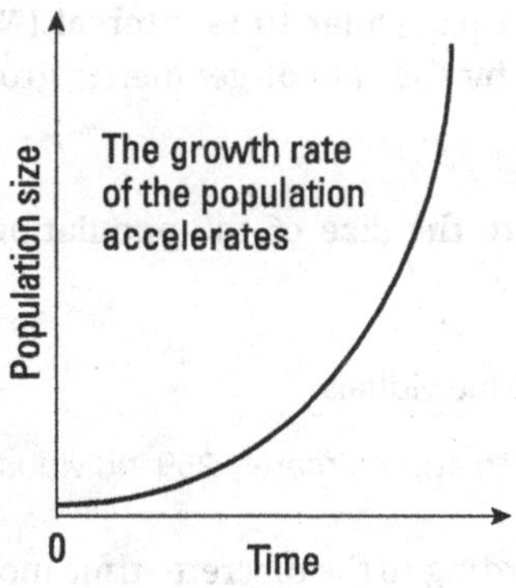

FIGURE 6-10: Exponential growth model.

The graph in Figure 6-10 shows exponential growth in a *continuous time model*, where reproduction happens continuously during the time interval. This model works for populations like humans and bacteria that reproduce all the time. Many other species only reproduce at certain times of the year, however. The population growth of those species is better represented by a *discrete-time model* that focuses on population time during a specific time interval.

The equation for exponential, or geometric, growth over a discrete time is:

$$N_{(t+1)} = \lambda N_t$$

This equation says that the population size for a specific time period (t+1) is equal to the starting population size (N_t) multiplied by the *rate of geometric growth* (λ). You can calculate the rate of geometric growth from known population sizes:

$$\lambda = N_{(t+1)} / N_t$$

For example, for the population described in the previous section that had 100 individuals and grew at a rate of 10 percent per year:

1. The starting population N_t = 100 individuals.
2. The population one year later $N_{(t+1)}$ = 110 individuals.
3. The rate of geometric growth for one year = 110/100 = 1.1.

If this population has a consistent growth rate each season, you can use this information to predict how the population will grow in the future. For example, if you wanted to know the population size after two years, you could take the population size at the end of one year and multiply it by the rate of geometric growth: 110 individuals multiplied by 1.1 give you 121 individuals at the end of year two. You could continue these hand calculations, or you can use this formula:

$$N_t = N_0 \lambda^t$$

In other words, the population at a particular time interval (N_t) is equal to the original population (N_0) multiplied by the rate of geometric growth raised to the power of the time interval (λ^t).

If you use this equation to calculate the size of the population above after ten years, it would look like this:

1. The original population (N_0) = 100 individuals.
2. 100 individuals multiplied by 1.1^{10} = approximately 259 individuals.

Unrestricted geometric growth according to the discrete-time model also increases exponentially, but the function on the graph is represented by discrete time points rather than a continuous line. Figure 6-10 illustrates the difference between the discrete-time model and the exponential model.

The logistic growth model equation is similar to the equation for exponential growth, but it adds a limiting factor:

$$dN/dt = rN^*(1-N/K)$$

This equation says that population growth rate (dN/dt) is equal to the per capita growth rate (r) multiplied by population size, *but also* multiplied by a limiting factor of the population size divided by carrying capacity, subtracted from the number one.

REMEMBER

As population size (N) gets close to carrying capacity (K), the value of N/K approaches one. When N/K equals one, then the limiting factor (1–N/K) equals zero, and the population growth rate becomes zero.

IN THIS CHAPTER

» Looking at the long-term impacts of species interactions

» Exploring relationships between predators and prey

» Competing to get what you need to survive

Chapter 7 Examining Interactions between Species

All living things engage in a web of interactions with other living things. You eat plants or animals, then microbes that live in your gut help you digest the food, producing molecules that affect your health and mood. *Biotic interactions* (*bios*=life) like these have short- and long-term impacts on individuals and populations. Your gut microbes have an immediate impact on your ability to digest food. Over long periods of time, humans and the bacteria that live in the human digestive tract have influenced each other in many ways, ultimately arriving at a mutually beneficial relationship where humans provide a home and bacteria provide molecules important to human health.

REMEMBER

Interactions between species influence populations and their evolution by natural selection.

Interactions that have negative impacts for individuals also influence populations. Predators limit the growth of prey populations and may change their behavior, causing them to move to more sheltered areas. Prey with traits that help them evade predators are more likely to survive and reproduce, passing defensive traits onto their offspring. Thus, over long periods of time, predators can drive evolutionary change within a species (see Chapter 3 for more on evolution and biological fitness). In this chapter, you'll explore the different types of biotic interactions and consider how those interactions may impact the survival of individuals and the evolution of populations.

Defining Interactions

Ecologists try to understand the impacts of biotic interactions on individual organisms as a way to determine how those interactions might contribute to each organism's success. They also try to predict how the interactions might shape the entire community (see Chapter 8 for more on communities), especially under changing environmental conditions. Some interactions seem obvious: If a wolf eats a rabbit, it's clearly a win for the wolf and a loss for the rabbit. Other interactions are harder to decipher, like a cow that grazes on grass but doesn't kill the grass. The cow did some harm to the grass, but maybe not enough to really impact growth and reproduction. And the cow might benefit the grass by eating invasive species that grow around it. Ecologists study these types of interactions both in the field and in the lab to determine how each species impacts the other.

When you consider all the types of impacts that one species could have on another, six possible results emerge, as shown in Figure 7-1. The figure shows all the combinations of positive (+), neutral (0), and negative (-) impacts, along with the terms that ecologists use to describe these interactions:

- In *mutualism*, both species benefit.
- In *commensalism*, one species benefits while the other is unaffected.
- In *antagonistic* interactions such as predation, herbivory, and parasitism, one species benefits at the other's expense.
- In *ammensalism*, one species is negatively impacted while the other is unaffected.
- In *competition*, both species are negatively impacted by their association.
- If neither species is affected by their association, the interaction can be called *neutralism*.

REMEMBER

The relationship between two species isn't always the same. Interactions can shift in response to environmental changes.

The relationship between a small gray moth called *Graya politella* and a wildflower known as the woodland star or *Lithophragma parviflorum* changes depending upon the environment. The moth is an important pollinator of the wildflower, receiving nectar as a reward. When the moth pollinates the flower, it also lays its eggs within the plant's ovaries. After the moth eggs hatch, the larvae feed on the plant's seeds. So, the moth always benefits from the interaction because it gets food in the form of nectar and seeds. What the plant gets out of the arrangement seems to depend upon the habitat of the plant. When ecologists studied the interactions between these two species over a range of environments, they saw three different types of relationships:

- In some areas, the relationship appeared to be mutualism. Seed production in pollinated flowers was higher than that in unpollinated flowers, so the plant benefited from its interaction with the moth.
- In areas that had plenty of other pollinators, the relationship seemed to be commensal. The moth still benefitted but didn't have a significant impact on seed development in the flowers.
- In a third group of study sites, the relationship between moth and wildflower was antagonistic, with the plant defending itself against the moth's parasitic larvae by aborting seed production in infected flowers.

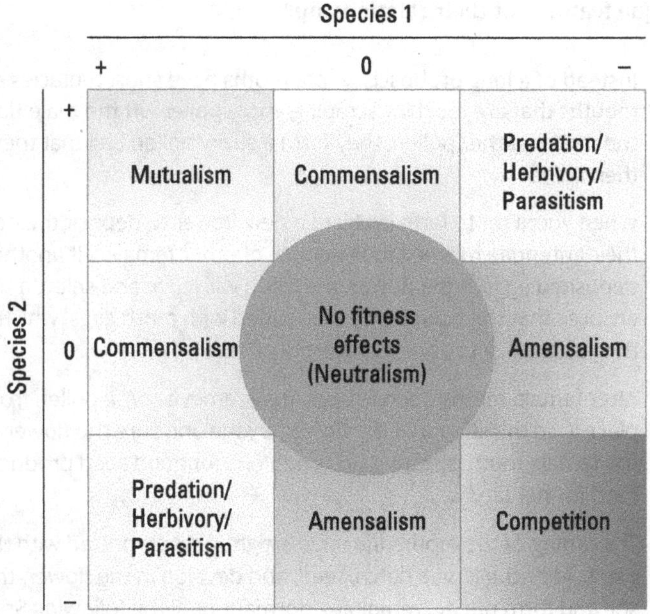

FIGURE 7-1: Types of species interactions.

Shaping evolution

Biotic interactions affect individual success, leading to changes in populations and community structure (see Chapter 8 for more on communities). Interactions like a predator capturing prey may be short-term, but the repeated influence of a predator population on a prey population can span long periods of time. Because biotic interactions contribute to biological fitness, they have the potential to shape the evolution of populations over time (see Chapter 3 for more on evolution).

We can see evidence today of the impact of biotic interactions over time in the close partnerships that exist between certain species, like that between yucca plants and their pollinators, the yucca moth. These two organisms have influenced each other's evolution to the point that neither can survive without the other.

REMEMBER

Coevolution occurs when two or more species reciprocally influence the evolution of the other.

Ecologists think that the relationship between yucca and its moth began as a simple feeding interaction, much like that between other flowers and moth pollinators. Typically, moths use their long, tongue-like proboscises to drink nectar from flowers, picking up and depositing pollen as they move from flower to flower. The results of coevolution between the yucca and its moth are apparent in several unique features of their relationship:

- Instead of a long proboscis, yucca moths have short tentacles around their mouths that are ideal for scraping yucca pollen off the male flower parts. As the moths gather pollen, they form a sticky pollen ball that they tuck under their chins.
- When yucca moth females fly to a new flower to deposit their eggs, they use their antennae to smell for evidence of other females. If another female has deposited eggs in the flower, the moth will leave and select a new flower. This ensures that the flower isn't overloaded with moth eggs, which might trigger the yucca plant to abort seed production.
- After female moths deposit eggs, they remove some pollen from their ball and place it on the stigma of the flower. By making sure the flower is pollinated, the female moth is setting up conditions for good seed production and lots of food for her larvae.
- The timing of the moth's life cycle is tightly synchronized with that of the yucca plant. After the larvae hatch, feed, and develop in the flower, they drop to the soil and form pupae, remaining dormant until the following Spring when the yucca blooms again. They emerge as adults and fly into the flowers to look for mates and begin the cycle again.

The tight partnership between yucca and their moths shows the result of coevolution over at least 40 million years. To understand how these species shaped each other in the past, scientists today investigate the genetic changes underlying the adaptations in each species. Another approach to understanding coevolution is to study variations in interactions that occur today, like the ones between the small gray moth and the woodland star described at the beginning of this chapter. The ecologists who study this system provide a window into coevolutionary processes as they happen in real time.

Meeting some mutualists

Ecologists consider the interaction between yucca plants and their moths to be a mutualism because the plant gets assistance with pollination and the moth gets food for its larvae. This is an example of a *dispersive mutualism*, in which one organism receives a reward for helping the other transport something like pollen or seeds.

Another type of mutualism is *trophic mutualism, where* two organisms help each other acquire food or other nutrients (*troph*=food). A classic example is the interaction between cows and the bacteria that live in their stomachs. Cows feed primarily on high-fiber plant material, but they cannot digest the fiber themselves. The bacteria in the cow's stomach can digest the fiber, also known as cellulose, and release some of the resulting glucose to the cow. As the cows graze, they provide a steady stream of food to the bacteria.

Some organisms engage in *defensive mutualisms*, where one species provides food or shelter in return for defense. Sea anemones provide shelter for clownfish, keeping predatory fish at bay with their stinging tentacles. In return, clownfish aggressively defend sea anemones against grazing by butterflyfish, who like to snack on the anemones' tentacles. Clownfish also help anemones by keeping them clean and improving water circulation.

Prowling with predators

Predators kill and eat other organisms. One way to think about predators is by what types of organisms, or *prey*, they eat:

- ***Carnivores* eat animal tissue.** Sharks tend to eat whatever animals they encounter, including fish, crustaceans, mollusks, marine mammals, and even other sharks.
- ***Herbivores* eat plant tissue.** White-tailed deer eat a variety of plants, including leaves from woody plants, green plants, seeds, and grass.
- ***Omnivores* eat plant and animal tissue.** Many people are omnivores, as are grizzly bears, which dine on salmon, berries, plants, insects, mammals, and carcasses that they scavenge from other predators.

Predators have obvious, direct negative impacts on the prey they capture, but their actions have other long-term impacts as well. When they kill an individual, they prevent that individual from passing its genes to the next generation, which can ultimately shape the evolution of the population. Predation also decreases the abundance and distribution of the prey population.

A successful predator must accomplish several components of predation:

- **Searching for prey:** As prey density increases, the time to locate prey decreases.
- **Attacking prey:** Most predators have adaptations that enable them to efficiently attack their prey.
- **Handling and digesting prey:** The amount of *handling time* a predator needs to manipulate and actually eat its prey is consistent and doesn't change with prey density.

Predators tend to look for food in areas that have abundant prey. Birds will travel farther from the nest to reach areas that have more food. Salmon sharks travel thousands of miles to reach spawning Pacific salmon. Even stationary predators like spiders and antlions preferentially choose locations with higher prey availability before they build their traps.

REMEMBER

Sit-and-wait predators stay in one place and attack prey that come within striking distance or enter a trap.

Optimal foraging theory

Foraging or building traps for prey requires energy from the predator. Since the goal of capturing prey is to gain energy, it makes sense that predators would want to gain more energy than they spend hunting or building. In terms of evolution, natural selection would favor predators with the highest energy efficiency because they'd be able to contribute more energy to reproduction. Ecologists apply this logic to the study of predator behavior.

REMEMBER

Optimal foraging theory helps ecologists make predictions about the foraging behavior of predators.

This theory considers the costs and benefits of foraging for particular prey as a way to predict which available prey represents the most profit, in terms of net gain of energy, to a predator. The model is based on several assumptions:

- All predators feed on certain types of prey.
- Each prey type has associated benefits and costs to the predator:
  - Benefits are the energy and nutrients they provide.

- Costs include how much energy they take to digest, how long the predator must spend hunting, and how much danger the predator must face.

» The energy profitability of each prey can be ranked in terms of the relative cost-benefit ratio.

Predictions made based on optimal foraging theory can be difficult to test in nature because capturing predators to assess their growth would disrupt their normal behavior. In one experiment with hummingbirds, ecologists solved this problem by installing perches into hummingbird territory that would record the weight of hummingbirds when they landed. When the ecologists covered half the flowers in the territory to make them inaccessible for nectar gathering, the hummingbirds responded by increasing their territory size. The hummingbirds seemed to optimize their territory size by trial and error until their daily weight gain reached a maximum value, suggesting that the hummingbirds were optimizing their energy profit.

One extension of optimal foraging theory is the *marginal value theorem*, which predicts foraging behavior in an environment where prey is scattered in patches separated by areas that lack prey. This theorem is represented in Figure 7-2:

» The curved line shows the total amount of energy captured by a predator after they arrive at a new patch:

  - At first, the amount of energy captured increases steadily as the predator captures more prey.
  - Eventually, the amount of energy captured slows as the prey resource is depleted. When the average available energy in the patch matches the average energy in the surroundings, the patch is no longer the optimal place to forage.

» As the amount of energy captured declines, the foraging is no longer optimal, so the model predicts the predator will leave the patch and seek another. The moment of departure is sometimes called the giving-up time.

  - The optimal time to leave the patch considers the amount of energy that it will cost the predator to travel to a new patch.
  - The model predicts the optimal time for the predator to leave the patch by calculating the tangent line from the travel time to the apex of the resource curve. (The travel time to the next patch is estimated based on the time it took to travel from the last patch to the current one.)

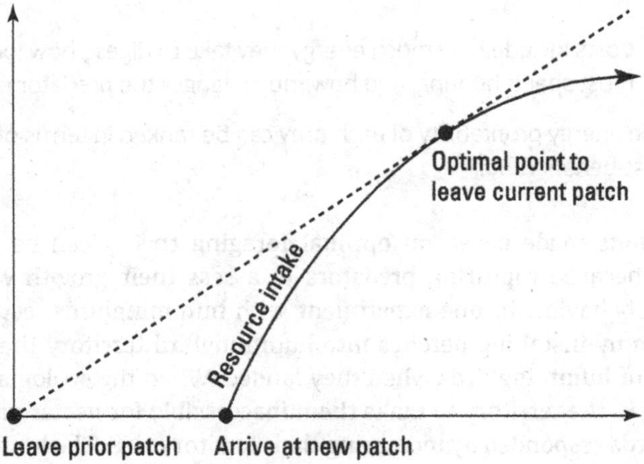

FIGURE 7-2: Marginal value theorem.

Studies on the foraging behavior of lizards, birds, and primates in nature have provided some support for the marginal value theorem. Recently, plant ecologists applied the theorem to the study of plant root growth by growing plants in three soil environments, each of which contained the same total amount of nutrients but distributed the nutrients in patches. Roots are a foraging organ, so the ecologists predicted that plants would invest more in their roots growing in highly fertilized patches of soil than those growing in low-quality patches. The mass of the roots in high-quality patches was greater than that of roots grown in low-quality patches, supporting the idea that plants minimize the costs of root growth to optimize resource acquisition. The plant roots in low-quality patches grew outwards from their patches sooner than did roots in high-quality patches, supporting the predictions of the marginal value theorem.

Risk-sensitive foraging

Many organisms find themselves in the middle of the food web, sometimes acting as predators and sometimes becoming prey. If you aren't a top predator, then going out to look for dinner can be hazardous to your health.

REMEMBER

Risk-sensitive foraging describes the behavior of animals that adjust their feeding behavior based on the perceived risk of becoming prey.

The presence of predators near a high-quality feeding area can reduce its value to an animal, causing it to forage in safer, lower-quality areas. Many lab and field studies support this idea, showing that foragers will shift their movement patterns and habitat selection in response to predator threat. For example, voles were more likely to forage in areas that did not have traces of weasel than those that did. Likewise, both mayflies and snails avoided areas that had chemical traces of their predators, leading to reduced foraging and reproductive success.

Ecologists have even conducted studies to determine how much incentive is enough to lure a forager into a dangerous area. Minnows were given the choice of two foraging areas, one with a single predator fish and one with two predator fish. When food was equal in the two areas, minnows foraged in the safer area. The ecologists gradually increased the amount of food in the danger zone and found that minnows would switch only when the food became twice as abundant. If more predators were added to the dangerous area, the minnows wouldn't switch until the food reached four times the original amount.

Predator-prey cycles

Predators and prey have direct impacts on each other's populations. Predators feed on prey, reducing the size of their populations. Likewise, the abundance of prey limits how many predators an ecosystem can support. Predators that have many prey options can switch between prey species in response to availability, but predators that have strong prey preferences are more impacted by the growth and decline of their prey populations.

The relationship between Canada lynxes and snowshoe hares shows how a tightly coupled predator-prey relationship can influence the population growth of each species. Canada lynx prefer snowshoe hares, who are mostly preyed upon by Canada lynx, so each species has a significant impact on the population of the other. This influence was first seen in the 1920s when ecologists graphed data from fur trappers that had been collected by the Hudson Bay Company during the late 19th and early 20th centuries. The number of pelts collected by hunters, shown in Figure 7-3, was considered an indirect measure of population size:

- » **When snowshoe hare populations increase, Canada lynx populations also increase after a short delay.** As more prey become available, lynx have more energy for reproduction.
- » **As Canada lynx populations rise, snowshoe hare populations start to decline.** Lynx capture and kill hares, removing them from the population.

REMEMEER

Predator-prey cycles, also called *population cycles*, are cyclical fluctuations in population numbers.

TECHNICAL STUFF

Two ecologists, Lotka and Volterra, created a set of mathematical equations to show how the interaction of predator and prey could result in population cycling. According to the *Lotka-Volterra predator-prey model*, the rate of change in the prey population is equal to the intrinsic growth rate of the prey population minus the removal of prey individuals by predators. In equation form, the model looks like this:

$$dR/dt = rR - cRP$$

The growth rate of the prey population (dR/dt) is directly proportional to the density of the prey population (R), multiplied by the maximum per capita growth rate of the prey (r) minus predation efficiency (c) multiplied by the density of the prey (R) and predator (P) populations.

Notice that if there aren't any predators, so that $P=0$, then the equation becomes $dR/dt = rR$, which is the same as the equation for exponential growth from Chapter 6 ($dN/dt = rN$). In other words, if there aren't any predators, this model predicts that prey will grow exponentially. The factors after the minus sign represent the loss from the prey population due to predation. As the population density of predators (P) increases, the number after the subtraction symbol gets larger, which means the growth rate of the prey population gets smaller. So, as predator population increases, the number of prey decreases, which is very similar to the pattern in Figure 7-3.

The other half of the Lotka-Volterra model predicts the impact of prey on predators with this equation:

$$dP/dt = acRP - dP$$

The growth rate of the predator population (dP/dt) is directly proportional to the predator birth rate (acRP) minus predator death rate (dP). Predator birth rate (acRP) is the number of prey captured (cRP) multiplied by a constant (a) that reflects how well the predator converts food into population growth. The per capita death rate (d) is multiplied by predator population size (P) to calculate the death rate (dP).

Notice that if the number of prey (R) is zero, then the predator population will decline exponentially as a factor of its per capita death rate (d) multiplied by the predator population size (P). In other words, without prey, the predators starve to death. As the prey population (P) increases, the number in front of the minus sign gets bigger, so the death rate has less impact on the total population size. So, as the prey population increases, the predator population increases, which again is very similar to the pattern in Figure 7-3.

The Lotka-Volterra model shows that changes in two factors, prey population size and predator population size, can explain the population cycles sometimes seen in nature.

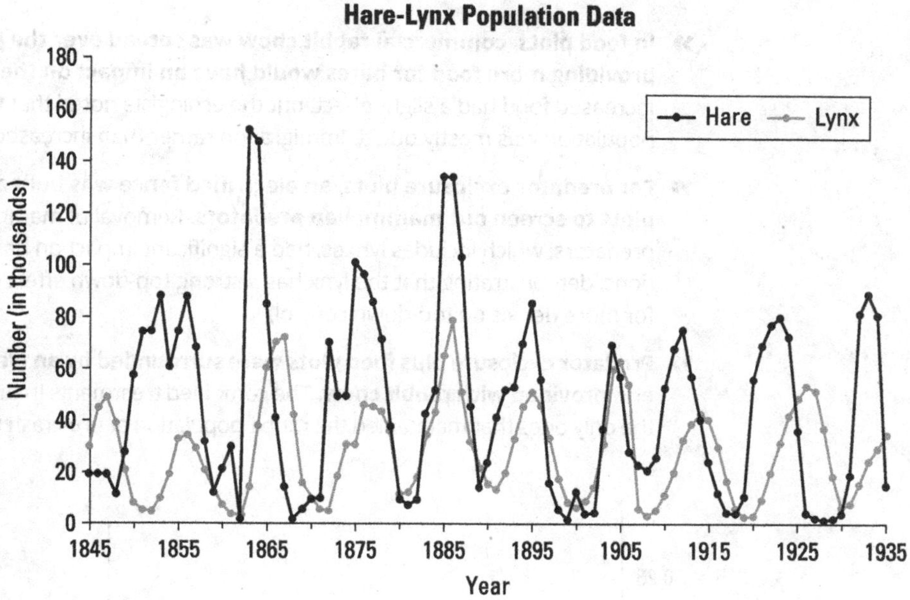

FIGURE 7-3: Population data on snowshoe hares and Canadian lynx are represented by the number of pelts collected by the Hudson Bay Company.

In nature population cycles can also be influenced by other factors, including the availability of food for prey, the presence of more than one acceptable prey species, or the presence of other predators. Ecologist Charles Krebs and his colleagues set up experiments to see how much other factors like these might affect the cycles seen in Canadian lynx and snowshoe hare populations. The ecologists marked off identically sized plots in the Canadian Yukon, captured snowshoe hares and fitted them with radio collars so they could track survival rates. Figure 7-4 shows snowshoe hare survival rates under different experimental conditions during a two-year period when the hare population was in decline:

- **In control plots, no treatments were administered.** These plots show the unaltered survival rate of snowshoe hares over the two-year period.
- **In fertilizer plots, commercial fertilizer containing nitrogen, phosphorus, and potassium was applied to the plots to test whether plant growth might control hare populations from the bottom up (for more details on bottom-up control, see Chapter 8).** The application of fertilizer had a slight but insignificant effect on the hare population, suggesting that plant growth was not a major factor in the population cycle.

- **In food plots, commercial rabbit chow was spread over the plots to see if providing more food for hares would have an impact on their population.** Increased food had a slight effect, but the ecologists noted that the increased population was mostly due to immigration rather than increased reproduction.
- **For predator exclosure plots, an electrified fence was built around the plots to screen out mammalian predators.** Removal of mammalian predators, which includes lynxes, had a significant impact on hare populations, demonstrating that the lynx has a strong top-down effect (see Chapter 8 for more details on top-down control).
- **Predator exclosure plus food plots were surrounded by an electrified fence and provided with rabbit chow.** The combined treatments in these plots were the only ones that maintained the rabbit population at pre-crash levels.

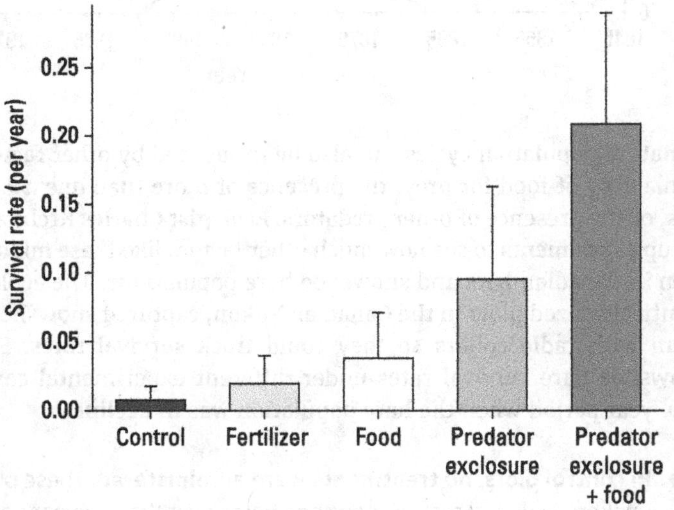

FIGURE 7-4: Annual survival rates of snowshoe hares from Fall 1990 to Fall 1992.

Taken together, these experiments showed that predation had the strongest impact on snowshoe hare populations. One thing that still puzzled the ecologists was why the predator exclosure plus food seemed to affect the hare population more than they would predict from the results of the separate treatments of added food or predator exclosure. It seemed that something about removing predators might increase the benefit of adding food to the enclosure. The scientists decided to look into an additional hypothesis that somehow the predators might be affecting the hares through sheer stress. You can read about the impact of stress on hares and other prey species in the sidebar "The Ecology of Fear."

THE ECOLOGY OF FEAR

Have you ever been so anxious that your stomach felt queasy and you didn't want to eat? Or had an experience so frightening that you developed an aversion to a particular location or even decided to stay home rather than go out? It turns out that these types of feelings and behaviors aren't unique to humans. Prey species that live under the threat of predators, like snowshoe hares living near Canadian lynx, may also exhibit these behaviors.

One of the clues that this might be true came from the predator-prey cycle, like that shown in Figure 7-3. Ecologists noticed that prey reproduction often starts to crash before the prey population reaches its peak, suggesting the decline starts before predator impact is at its greatest. When ecologists tested the blood and feces of hares for chemicals associated with stress, they found that breeding females had the highest stress levels during summers of peak predation and that they produced fewer, smaller offspring. Their offspring also had higher stress levels and produced fewer offspring than hares not under threat of predation. These results helped explain why prey reproduction remained low even after predator numbers fell.

Experiments like these support the idea that predators have indirect negative impacts on their prey populations. Ecologists studying small mammals and songbirds have also found that fear of predators causes these animals to have fewer offspring. Signs of behavioral changes as a result of trauma may also occur in larger animals, such as elephants. The fact that animals other than humans may experience lasting impacts from stress is causing some scientists to think differently about human responses, such as post-traumatic stress disorder (PTSD). It may be that PTSD has long evolutionary roots and is similar to adaptive survival strategies in other species. This way of looking at PTSD may open up new avenues of research.

Stabilizing predator and prey populations

Predator-prey cycles like the one between Canada lynx and snowshoe hares aren't very common in nature, occurring mostly in arctic ecosystems. Predator-prey cycling occurs when predators and prey have few options other than to eat or be eaten by the other. Although Canada lynx do migrate away from feeding areas once the hare population is low, the impact of predation on the hare reproduction continues after their departure, possibly because of the influence of stress (see the sidebar "The Ecology of Fear" for details). This time delay in the recovery of the population is part of the reason for the population cycles. In other environments, factors may have mitigating effects that reduce the high and low swings of predator-prey populations:

» **Predator inefficiency:** The ability of prey to defend themselves or escape from predators makes it harder for predators to hunt efficiently. If predators aren't as efficient at capturing prey, then increases in predator populations won't be as tightly coupled to decreases in prey populations.

- **Density-dependent limitations on predators or prey populations:** If factors outside their relationship limit predator and prey populations, then population numbers won't always rise exponentially.
- **Alternative food sources for predators:** If predators can switch to other prey when the population of one type gets low, then they won't hunt a single prey population to very low numbers.
- **Refuges where prey can hide themselves from predators:** When prey can hide, their populations survive at higher levels, allowing them to recover more quickly.

Predators can reach a point where they are *satiated*, or have eaten enough food, and will stop hunting. This is more likely to happen when prey are abundant. Also, when prey populations are low, it may be harder for predators to locate prey. Thus, prey abundance can impact feeding behavior in multiple ways.

REMEMBER

The *functional response* of a predator is the relationship between its rate of food consumption and the abundance of its prey.

Ecologists propose three basic types of functional responses, shown in Figure 7-5:

- **Type I functional response:** The number of prey eaten is directly proportional to prey density. This relationship rarely occurs in nature.
- **Type II functional response:** The number of prey eaten increases rapidly as prey density increases, but then levels off at high prey density. This response is predicted because the time to locate prey decreases as prey density increases, but handling time remains the same. So, predators can locate and kill more prey initially, but their ability to kill reaches a maximum because it takes a consistent amount of time to handle and eat their prey.
- **Type III functional response:** At low prey density, the number of prey consumed is lowered, but then accelerates as prey density begins to increase. This model reflects situations where prey have refugia, or places to hide. At low prey densities, all individuals may have access to the refugia and be difficult for predators to find. Additionally, when the density of one prey population is low, predators may switch to a different prey type. Then, as the prey population increases, more prey are exposed, and predators don't have to search as long. At high prey density, predators don't have to search very long, but their handling time limits their ability to hunt any faster.

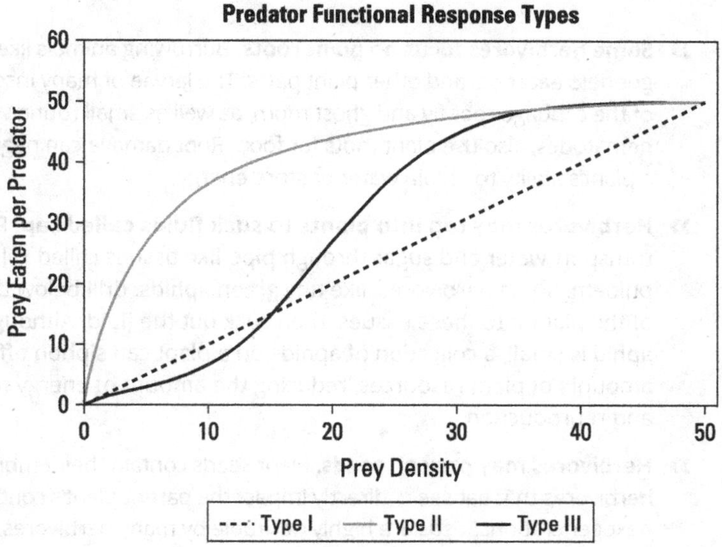

FIGURE 7-5: Types of predator functional responses.

Ecologists use functional response models to form hypotheses about ecosystem function. Once a particular functional response is identified in a system, ecologists can look for the mechanisms that cause the response. They can also investigate which types of responses help stabilize ecosystems.

Hanging out with herbivores

Herbivores can be considered predators if they eat whole plants or parasites if they just remove some plant parts. Some herbivores, like elephants, will eat many different types of plants and their parts, but others may specialize in just one plant and even just one part of that plant. Koalas, for example, survive almost exclusively on the leaves of eucalyptus trees. Insects, which are the most common herbivores, also tend to be specialists.

Herbivory typically has direct negative effects on plants, resulting from tissue loss. Depending upon what part of the plant is removed, the impact may be obvious and immediately visible, or it may be more subtle:

» **Most herbivores dine on plant leaves.** Larger herbivores such as deer will pull entire leaves of a plant, while smaller herbivores like caterpillars remove tissue from leaves one tiny bite at a time. Leaves are the primary photosynthetic organs of most plants, so their removal or damage directly impacts the plant's ability to obtain energy. This can reduce the amount of energy available for reproduction, lowering the overall fitness of the plant.

- **Some herbivores focus on plant roots.** Burrowing animals like moles and gophers eat roots and other plant parts. The larvae of many insects, like those of the cabbage root fly and ghost moth, as well as small round worms called nematodes, also use plant roots for food. Root damage can negatively impact a plant's ability to obtain water or store energy.
- **Herbivores may tap into plants to suck fluids called sap.** Plants transport water and sugar through pipe-like tissues called xylem and phloem. Some herbivores, like tiny green aphids, drill below the surface of the plant into these tissues, then suck out the fluid. Although each aphid is small, a collection of aphids on a plant can siphon off significant amounts of plant resources, reducing the amount of energy for growth and reproduction.
- **Herbivores may prey on seeds.** Plant seeds contain their embryos, so herbivores that eat seeds directly impact the parent plant's contribution to the next generation. Seeds are highly desirable by many herbivores, including birds, rodents, and insects, because they're packed with nutrition for the developing embryo. *Seedivores* are true predators because they consume entire organisms when they eat seeds.

The degree of the negative impact of herbivory on plants also depends on how much the herbivore consumes. If the herbivore doesn't eat too much of any one plant or population, or if the herbivore provides a resource like fertilizer from their waste, then the overall impact of the interaction might be more neutral. In comparison, overgrazing by livestock animals can decimate populations in grasslands, contribute to soil degradation and runoff, and ultimately cause shifts in community structure and ecosystem health. (See Chapter 8 for more on community structure.)

Competing with Your Neighbors

Competition in nature is similar to that between people in that individuals try to get something before another can take it. As organisms compete for resources like food, light, water, and space, they negatively impact each other. Competitive interactions may be indirect or involve direct action:

- ***Resource competition*, also called *exploitative* competition, happens when one individual uses a limited resource, preventing others from getting what they need.** In this type of competition, the negative impact results from indirect interaction. Trees growing too closely together may have overlapping leaves that block each other's access to sunlight, or a fast-growing species of barnacle might prevent a slower-growing species from surviving on an intertidal rock.

- ***Interference competition* occurs when one organism directly acts to block another from accessing a resource.** For example, hummingbirds will actively chase other birds away from food, and male gorillas use aggressive displays to prevent other males from getting mates. Some plants use chemical competition, called *allelopathy,* to stop other plants from growing near them. Black walnut trees produce a chemical called juglone that inhibits enzymes in other plants. The presence of juglone in the soil around black walnut trees prevents the seeds of other plants from germinating.

The classic experiment to determine whether two species are competing with each other is called a *common garden experiment.* This type of experiment got its name because it was first done in plants, but ecologists have since used it to test for competition in all kinds of species. The experiment in Figure 7-6 shows how each species can be grown alone and then with others to test for negative impacts on growth and reproduction from competition:

- Each species is grown alone as a control for growth and reproductive potential without competition. Various observations may be collected, including fundamentals like growth measurements, weight, and number of offspring.
- To test for *intraspecific competition*, multiple individuals of the same species are grown together (*intra*=within, so intraspecific=same species). These organisms are grown for the same amount of time as the controls, and the same measurements are collected.
- To test for *interspecific competition* individuals of different species are grown together (*inter*=between). The same number of individuals would be grown together as in the intraspecific treatments.

The growth of individuals grown with others is compared to the growth of individuals grown alone. If individuals in the intraspecific or interspecific treatments have reduced growth or reproduction when compared to control individuals, then competition is occurring. If one species is a stronger competitor than the other for a resource, then it will grow and reproduce better in interspecific treatments.

Competition between species can shift in response to environmental conditions. A plant species that is a better competitor than another in sandy soils might be unable to compete in nutrient-rich soils. Competition between plants for nitrogen and phosphorous may be eliminated by the application of fertilizer. A species that can normally outcompete another might fall behind in response to drought or temperature changes. The increasing global temperatures due to climate change (Chapter 12) have the potential to cause shifts in the interactions between species, changing community structure (Chapter 8) and ecosystem function.

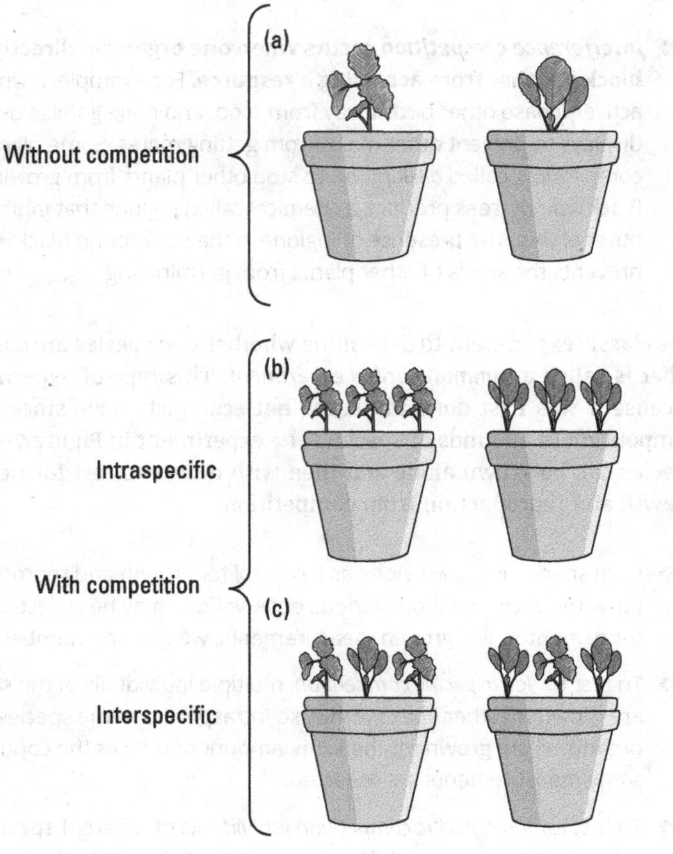

FIGURE 7-6: A common garden experiment to test for competition.

The presence of predators can affect interactions between other species in a way that looks like competition but isn't really about the ability to acquire resources. In *apparent competition*, the action of the predator reduces the population of a highly competitive species, allowing other species to grow. For example, grazing animals can cut back dominant grasses, allowing other grasses and wildflowers to grow. When species are moved from one environment to another due to human activity, they may no longer have predators to keep their populations in check. If they are able to resist predators in their new environment, they can reproduce rapidly and destabilize local ecosystems (for more about these invasive species, head to Chapter 9). These examples represent apparent competition because the action of the predator is controlling population growth, not the ability of the species to compete for resources.

Getting what you need to survive

Resources are anything that can be consumed and are needed to promote population growth. As a resource declines in abundance, population growth declines with it.

Examples of resources include food, water, light, space, and refugia. Some resources can be renewed, others can't:

- *Nonrenewable resources* aren't regenerated. Once an organism occupies a space, that space is unavailable unless the consumer of the space leaves.
- *Renewable resources* are constantly replenished. Prey organisms reproduce, creating more food for predators. Decomposition of dead organisms recycles nutrients for others to reuse (for more on matter cycling, flip to Chapter 11).

Some resources may be so rare in an environment that they limit the growth of a population. Plants and algae both require nitrogen and phosphorous to grow, but these elements may be rare in the environment. The addition of these elements through fertilizers can overcome the limitation and lead to rapid growth. When pollutants containing these elements enter waterways, sudden algal growth destabilizes the ecosystem, leading to eutrophication (see Chapter 10 for more on eutrophication).

REMEMBER

When a resource that's needed for population growth is insufficient in an environment, it acts as a *limiting resource*.

Competition increases when resources are scarce. If two species use the same limiting resource in the same way, the better competitor will ultimately drive the other species out.

REMEMBER

The *competitive exclusion principle* states that two species can't coexist indefinitely when the same resource limits both species.

Species that are very similar in their requirements can sometimes share a limiting resource through *resource partitioning*, where each species shifts its use of the resource slightly:

- **Species may use the resource at different times:** Hairy woodpeckers and black-backed woodpeckers both prefer to forage in lodgepole pines with very similar characteristics, but they fledge their young at different times, reducing the potential for competition.
- **Species may use different areas of the resource:** Wood warblers are colorful, insect-eating songbirds. Wood warblers that coexist in the same forests forage in different areas of the same trees, some foraging near the ground while others hover and forage at different positions in the tree. Although they eat similar diets, so they are using the same food resources, they forage in different locations, which reduces competition.

- **Species may use different fractions of the resource:** Large mammals in Africa, like zebras, dik-diks, and impalas, can coexist in the same savannah because each species eats slightly different plant species. Zebras graze on grass, dik-dik's browse on larger plants, and impalas graze and browse. Each species uses a different fraction of the available producers, which reduces competition and allows them to coexist.

Figuring out your niche

The idea that different species can appear to need very similar resources, but actually have unique specializations that make them distinct in their requirements, is wrapped up in an idea called the *niche concept*. You can think of an organism's niche as its lifestyle or job description. The niche spells out exactly how a species fits like a puzzle piece into the larger picture of its ecosystem.

REMEMBER

The *niche* of a species includes all aspects of its habitat, how it makes a living, and the physical environment in which it is found.

Different species of woodpeckers, or wood warblers, or African mammals may appear to be very similar at first glance. Still, when you carefully examine how and where they get their resources, you identify the differences that define their niches. Although they have many similarities, they occupy different niches, which is why they can coexist.

REMEMBER

Another way of stating the competitive exclusion principle is to say that two species can't coexist in the same niche.

The niche concept can be divided into two parts:

- The range of physical conditions a species could survive in is its *fundamental niche*.
- The actual distribution of a species is its *realized niche*.

REMEMBER

The realized niche of a species may be smaller than its fundamental niche because of interactions with predators, pathogens, and competitors.

Living Together

Sometimes, interactions between organisms become very intimate and long-lasting. Most plants have associations with soil fungi called mycorrhizae that last their entire lives. Healthy corals contain single-celled photosynthetic organisms that live inside their bodies. You received microbes from your mother at birth, and the descendants of those microbes continue to live in and on your body today.

REMEMBER

Symbiosis occurs when two organisms live together for a significant portion of their life cycle.

The term symbiosis can mean different things to different people. But in the strictest use that goes back to the origin of the word (*sym*=together, *bios*=life), it refers to significant, long-lasting interactions between different species.

WARNING

In common usage, people sometimes use the term symbiosis to mean a good thing, but symbiosis covers all possible outcomes of a long-term interaction, from mutual benefit to parasitism. The key to whether a relationship is symbiotic is how long the interaction lasts, not the outcome. (Some companies adopted the word symbiosis in their names, and I don't think they mean to call themselves parasites.)

It can be very difficult to determine exactly how the partners in a symbiotic relationship affect each other. In the ideal test, you would grow each partner independently and measure their growth and reproduction. Then, you'd compare those measurements to growth and reproduction rates when the partners are together. This would let you determine whether the partners are better together or if one or both actually grow better on their own. One difficulty here is that some symbiotic partners can't be grown separately, at least not easily. Scientists who study symbiosis may spend long periods of time trying to figure out how to support independent growth and still not find a way.

REMEMBER

In many cases, scientists infer the impact of a symbiotic relationship on the two partners from indirect evidence, rather than experimental results.

Friends with benefits

If you want to refer to a win-win symbiotic relationship, you call it a *mutualism* or a *mutualistic symbiosis*. As with more casual mutualistic relationships, both partners benefit from this type of symbiosis.

Lichens are one of the most well-known examples of a mutualistic symbiotic relationship. Lichens range from colorful crusts on rocks that almost look like paint splotches to leafier growths that can easily be mistaken for moss or plants. Lichens are actually unique partnerships between fungi and photosynthetic organisms like single-celled algae. The algae do photosynthesis, providing the fungus with sugar. The lichen body, built by the fungus, protects the algae from dying from too much sun or a lack of water. Together, the partnership can grow in places where neither one could survive alone.

Symbiotic relationships flourish in the digestive systems of animals. Many of these are mutualistic partnerships where the host provides food and a nice place to live, and microbes help with food digestion. Ruminant animals like cows could not get enough nutrition by grazing if it weren't for their microbial partners that can break down cellulose in plant material. When the microbes digest the cellulose, some of the resulting glucose becomes available to the animal host. Likewise, termites can't digest wood on their own. The breakdown of resistant molecules in the wood occurs because of bacteria that live inside single-celled protozoans that live inside the termite hindgut!

Some microbes that live in animal digestive systems may not actually participate in the digestion of food molecules or produce any molecules useful to the host. The relationship between these microbes and their host could be described as a commensal symbiosis because the microbes benefit, but the host gets nothing in return. Another example comes from the ocean, from tiny little goby fish. These fish are masters of camouflage, changing their colors so they can hide on the surfaces of other species like corals, algae, and sea sponges. The gobies get a place to hide, but they probably don't affect the organism that's sheltering them.

Parasites and free-loaders

Symbiosis has a darker side, one that includes organisms that improve their own fitness by harming others. Unlike the quick and deadly attack of a predator, parasites and parasitoids prefer the long, slow drain of their host organism.

REMEMBER

Parasites take nutrients and energy from their hosts without immediately killing them.

Some parasites, like the fleas that live on dogs and cats, never kill their host. Others, like the eukaryotic microbes that cause malaria, can be deadly.

The life cycle of a malarial parasite, shown in Figure 7-7, shows just how complex the relationship between parasites and their hosts can become. In order to complete its life cycle, the parasite must spend time in two different hosts, humans and mosquitoes. In each host, the parasite changes forms as it moves through different tissues:

1. Pregnant mosquitoes seek out a blood meal to feed their developing eggs. If the female mosquito is infected with the malarial parasite, they inject one form of the parasite, called a *sporozoite*, into the blood as they feed.
2. Once in the blood, the parasite migrates to the liver, where it enters liver cells.

3. Inside the liver cells, the parasite reproduces asexually to form thousands of another cell type called a *merozoite.*
4. The merozoites swarm out of the liver, some taking on another form that infects red blood cells. The parasite continues to reproduce, destroying red blood cells in the process.
5. Other merozoites begin the process of sexual reproduction, producing cells that will become male and female gametes.
6. When a mosquito bites an infected person, they ingest these sexual cells. Once inside the mosquito's gut, the cells finish developing into sperm and egg cells.
7. When a sperm cell fuses with an egg cell, they form yet another form of the parasite, which glides through the mosquito until it finds the salivary glands. Once there, these cells produce new sporozoites, continuing the life cycle.

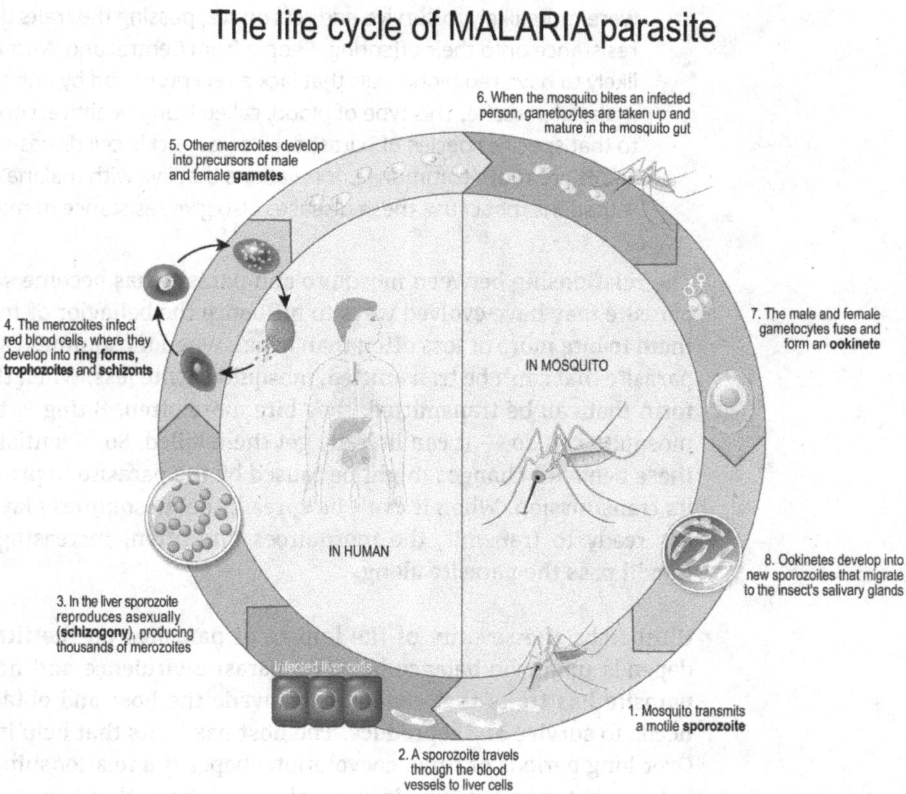

FIGURE 7-7: The life cycle of a malarial parasite.

designua/Adobe Stock Photos

Malarial parasites have symbiotic relationships with two hosts, requiring both hosts to reproduce sexually and complete their life cycle. Over time, the parasites and hosts have affected each other's fitness, causing adaptations in both parasite and host:

- **Parasite adaptation:** Malarial parasites infect a wide variety of species, including birds, rodents, monkeys, and humans. Each species of parasite is adapted to specialize in a certain range of hosts. Some species have adapted to the point that they can only infect humans and certain types of mosquitoes. In one experiment, parasites that normally reproduce only in anopheline mosquitoes were introduced into *Culex* mosquitoes. The parasites died in the mosquito gut, perhaps because they were recognized and destroyed by the *Culex* immune system. This suggests that the parasite has evolved ways to avoid immune detection in anopheline mosquitoes.
- **Host adaptations:** The malarial parasite shaped the evolution of human populations living under its threat. Individuals with some form of resistance were more likely to survive and reproduce, passing the traits that gave them resistance onto their offspring. People from Central and West Africa are more likely to have red blood cells that lack a receptor used by one species of malarial parasite. This type of blood, called Duffy negative, confers resistance to that specific species of parasite. Likewise, sickle cell disease and thalassemia are more common in populations that live with malaria because the mutations that cause these diseases also give resistance to malarial parasites.

The relationship between mosquito and parasite has become so intimate that the parasite may have evolved ways to influence the behavior of mosquitoes, causing them to bite more or less often than usual. When mosquitoes carry the form of the parasite that can't be transmitted, mosquitoes bite less. When they're carrying the form that can be transmitted, they bite more often. Biting is the riskiest thing a mosquito can do — it can literally get them killed. So, scientists hypothesize that these behavior changes might be caused by the parasite to protect it and enhance its transmission. When it can't be spread, the mosquitoes play it safer. But when it's ready to transmit, the mosquitoes bite often, increasing the chances that they'll pass the parasite along.

Ultimately, the severity of the impact of parasites on the fitness of their hosts depends upon the balance between parasite virulence and host resistance. The parasite has traits that enable it to invade the host and obtain the resources it needs to survive and reproduce. The host has traits that help it fight the parasite. Over long periods of time, coevolution shapes the relationship, often leading to a balance between the two. In general, populations that have lived with a parasite for long periods of time show greater resistance than populations that have never encountered the parasite before promoting tolerance. Usually, the greatest risk of death is when a parasite is first introduced into a new population, and the hosts have no resistance.

Depending upon the interplay between host resistance and parasite virulence, a parasite may or may not kill its host. With parasitoids, however, death of the host is always the outcome.

REMEMBER

Parasitoids have young that develop on or within the host, eventually killing it.

Parasitoids are insects like flies, wasps, and beetles. Like the saber wasp shown in Figure 7-8, parasitoids lay their eggs in or near host species. When the female lays her eggs, she often injects the host with toxins that cause paralysis or arrest development, making it easier for her larvae to feed after they hatch. The developing young feed carefully, keeping the host alive until they are ready to complete their own development.

Parasitoids shape community structure because they exert top-down control of other insect populations (see Chapter 8 for more details on top-down control). They're beneficial to humans because they feed on crop, garden, and forest pests. Some farmers and forest managers use parasitoids for biological control of harmful species.

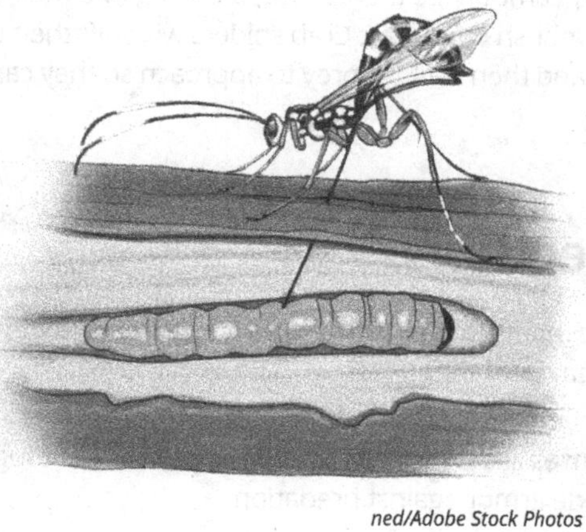

FIGURE 7-8: A parasitoid wasp lays eggs.

ned/Adobe Stock Photos

Adapting to Different Lifestyles

Interactions between species over long periods of time influence their evolution. The predators that are capturing prey are more likely to reproduce, passing on traits that made them successful to their offspring. Likewise, prey that have the best defenses might escape to live another day and pass on their skills. The results of this shifting dance between predator and prey are visible today in the adaptations of different species.

Predatory offensive skills

Most predators have evolved to be very effective at locating, attacking, eating, and digesting their preferred prey. Depending upon their specialization, they may have physical, chemical, and behavioral adaptations that assist them with these tasks:

- » Physical adaptations include body form, mouth structure, and the type of teeth. Predators built for bursts of speed, like the cheetah, can run down rapidly fleeing prey. The bones in a snake's skull are attached in a way that lets them move past each other so that the snake can open its mouth wide enough to eat large prey. The teeth of carnivores include sharp canines for tearing flesh, while the teeth of herbivores are large and flat for grinding fibrous plant material.
- » Predators may use chemicals for offense or defense. Venomous spiders paralyze their victims with their bite, making them easier to kill and eat. Some herbivores produce enzymes that break down plant defensive chemicals. Other herbivores have symbiotic bacteria in their digestive systems that perform this detoxification for them.
- » Predators may camouflage themselves, blending into their environment as they wait to ambush their prey. Crab spiders will shift their colors to match flower petals and then wait for prey to approach so they can grab them with their front legs.

Prey protections

Prey species have many adaptations to escape being eaten, including physical, chemical, and behavioral defenses:

- » Physical defenses like the spines on a porcupine or the tough shell of a tortoise provide armor against predation.
- » *Crypsis* or camouflage allows prey to hide in plain sight. Stick bugs may avoid detection by birds as long as they don't move too much. Likewise, leaf bugs look like leaves and have colors that let them blend into the background.
- » Some prey use *warning coloration* to advertise the fact that they contain toxic compounds. Black contrasted with bright red, orange, or yellow is a common color combination among these bad-tasting animals. When animals copy each other to advertise that they are noxious, ecologists call it *Müllerian mimicry*.

- *Batesian mimicry* occurs when harmless organisms copy the warning coloration of noxious species. A species of robber fly that adopted the bright yellow and black stripes of its prey, the bumblebee, may escape predation by toads.
- Some prey take actions such as *startling behavior* or playing dead to deter predators. Startling behavior attempts to scare off the predator with a sudden display, like the flash of large eye-shaped spots on the wings of a butterfly. Other animals use *distracting behavior,* like a cloud of ink from an octopus, to help them make their escape.

Plant defenses

Plants can't run away from predators, but that doesn't mean that they're defenseless. You've probably seen physical defenses like spines and thorns that can discourage browsing animals. In addition, plants produce powerful chemicals to defend themselves against their predators. Many of the flavors you may enjoy from plant foods, like the heat in chilis or the slightly bitter taste of coffee, come from plant defensive compounds.

REMEMBER

Plants make chemicals called *secondary compounds* for defense against herbivores and parasites.

Scientists call these chemicals secondary compounds to distinguish them from the primary compounds that plants use for basic functions like photosynthesis and growth. If a compound is produced all the time, it's *constitutive*. Compounds that plants produce in response to damage or other signals are *induced*.

TIP

Constitutive sounds like constitution, so constitutive defenses are part of a plant's basic health. Induced means to turn on, so induced defenses get turned on when they're needed.

For people, many plant secondary compounds are flavorful or medicinal. But for herbivores, they can be repellent or even poisonous. Caffeine, for example, is something many people rely on, but it can paralyze and poison insects, negatively impacting their ability to reproduce.

3 Forming Communities and Ecosystems

IN THIS PART . . .

Get along with the neighbors

Rise from the ashes

Go with the flow of energy

Learn to recycle as a way of life

IN THIS CHAPTER

» Embracing an ecological view of communities

» Delving into species diversity

» Meeting the biggest influencers in the community

Chapter 8

Engaging with Communities

How well do you know your community? Do you know who lives there? Not just the people, but the plants, the other animals, the mushrooms, and the microbes? Whether you're aware of them or not, all these types of organisms make up the web of living things around you. You might notice them when they directly interact with you; for example, if birds beat you to your own blueberries. But do you notice the more subtle, indirect interactions? You might not notice if neighborhood cats reduce the bird population, indirectly benefiting you with a better blueberry harvest. When ecologists consider communities, they try to understand all the interactions and how they affect the members of the community. In this chapter, you'll take a look at how ecologists define communities and examine the factors that have the power to change them.

Defining Communities

Organisms interact with each other in different ways, ranging from deadly predation to beneficial mutualism (flip back to Chapter 7 for details). It may be easy to understand individual interactions, for example, how wolves might impact elk populations in Yellowstone National Park. Wolves are predators of elk, so the obvious prediction is that wolves will lower the elk population. It's much harder to predict the outcome of that interaction when you consider it in the context of the interactions of all the organisms that form a community in Yellowstone National Park.

REMEMBER

Communities are groups of interacting species living together in a particular place at a particular time.

The Yellowstone community was completely transformed, first by the extermination of wolves in the 1920s, and then by their reintroduction in 1995. The ripples of impact extended from the wolves to the elk and other species, and even to the landscape. (For details, see the "Where the Wild Things Live" sidebar.)

WHERE THE WILD THINGS LIVE

In 1872, the United States government created its first national park. Because of its unique beauty and natural resources, Yellowstone National Park was protected from development and "set apart as a public park or pleasuring-ground for the benefit and enjoyment of the people." Initially, people could use the park to hunt game, but this practice ended in 1883 when Congress passed laws that prevented the hunting of park animals. However, the hunting ban didn't include protection for predatory species such as wolves. Local farmers saw wolves as a threat to their livestock, so the wolves were routinely hunted or killed by park officials, leading to the destruction of all the Yellowstone wolf packs by 1926.

The loss of the wolves triggered a cascade of change. The elk population exploded, leading to overgrazing of willow and aspen trees. Without the trees, the beaver population declined, and their dams disappeared, leading to the loss of small ponds and increasing erosion along streams. The lack of elk killed by wolves also meant there was less food for scavengers, including bears, eagles, and ravens. Coyotes became the dominant predators, reducing the populations of their prey species such as antelopes, red foxes, and rodents. The sum of these changes lowered the diversity of species living in Yellowstone and led to the degradation of the landscape due to erosion.

Scientists and environmentalists began advocating for the return of the wolves beginning in the 1940s. Still, they didn't have any legal means of making it happen until the Endangered Species Act was passed in 1973. This law required the US Fish and Wildlife Service to develop a plan to help wolf populations recover. Finally, in 1995, US and Canadian wildlife officials captured 14 Canadian grey wolves and introduced them to Yellowstone. To increase genetic diversity, 17 more wolves were added in 1996.

The wolves established themselves in the park and increased their population. Predation by wolves combined with hunting by humans has reduced the elk population. The presence of the wolves helps support grizzly bear populations because the grizzlies steal elk carcasses away from wolves. The decline in elk allowed recovery of some aspen and willow populations. The recovering forests attracted many species of birds and allowed the beavers to return. The beavers built their dams again, providing cool shaded areas for fish and creating new pools where species like otters and mink could thrive. When wolves hunt coyotes, they help mice, rabbits, and foxes survive. Although there's some debate among scientists about how much of these changes can be attributed to the reintroduction of wolves, it's clear that wolves have influenced all these community dynamics.

Community ecologists seek to understand how communities form and function as a unit. Two major hypotheses challenge the way ecologists think about communities:

- The *holistic concept* of communities suggests that the organisms in a community interact as functioning parts of the whole, much like the organs of the human body support its function. In this view, the populations in the community are interdependent, and the sum of all their interactions contributes to the success of the community as a whole.
- The *individualistic concept* of communities sees the populations within a community as more independent of each other. They interact because they all live together in the same place and time. They are capable of surviving within the community, but their individual actions determine their reproductive success.

Ecologists identify communities based on their physical and biological characteristics. Some communities have clearly defined boundaries and don't overlap much with neighboring communities, while others blend more gradually into their surroundings. For example, the community living in a strip of salt marsh between a tidal marsh and an upland grassland may have species in common with both of its neighbors. In contrast, distinct boundaries often occur in places where the environment changes rapidly along a gradient.

REMEMBER

An *ecotone* is a boundary between ecosystems.

Ecotones form along edges, like where water meets land. The community living in a mountain lake probably has very few species in common with the community occupying the surrounding land. Sometimes, the presence of a particular organism can determine the boundary of a community. For example, coral reef communities only exist where corals grow.

Ecologists who study communities often focus on subsets of species within the community:

- **An *assemblage* is a group of species that are related to each other.** Insect populations respond quickly to changes in habitat, so ecologists might study all the ground beetles in a geographical area as a way of assessing the impact of land use changes on local communities.
- **A *guild* is a group of species that use the same resources in the same way but aren't necessarily related to each other.** An ecologist might investigate the importance of pollinator diversity on crop yield by studying all the pollinators in an orchard, including wild bees, farmed honey bees, butterflies, fungus gnats, flies, and beetles.
- ***Trophic levels* group species based on how they obtain energy.** *Primary producers* take energy and matter from the environment to make food. *Primary consumers*, or *herbivores*, consume primary producers for matter and energy. *Secondary consumers*, or *carnivores*, consume herbivores. (Chapter 11 presents the energy relationships between organisms in more detail.)

Two of the most important factors that influence a community are the number of different species it contains and the specific identities of those organisms. These factors determine the characteristics that identify the community type or structure.

REMEMBER

Community structure refers to the set of characteristics that shapes the community.

Measuring species diversity

Species diversity considers the variation in a community on multiple scales, from genes to phylogenetic taxa (see Chapter 13 for details). Two factors that are commonly used are species richness and species evenness:

- *Species richness* is the number of species in the community.
- *Species evenness* is the relative abundance of individuals within each species. Abundance is measured in different ways depending upon the species:
  - The number of individuals within each species can be counted.
  - For plants, the percentage of the area occupied by each species can be measured. For example, one species might cover 25% of a meadow, while another occupies just 10%. These measures of abundance are reported as *percent cover*.
  - The relative weight, or *biomass*, of species may also be used to assess abundance.

Figure 8-1 compares two communities that have the same species richness but differ in their overall diversity. Both communities have the same number of species, represented by the unique symbols. If you count the number of each type of symbol, however, you'll see differences in the species evenness. In the community on the left, one species is much more common than the other two, so that the species aren't evenly represented. Another way of thinking about this community is that, although all species are present, some are rare. The community on the right has the same species richness, but greater evenness, so it is more diverse overall.

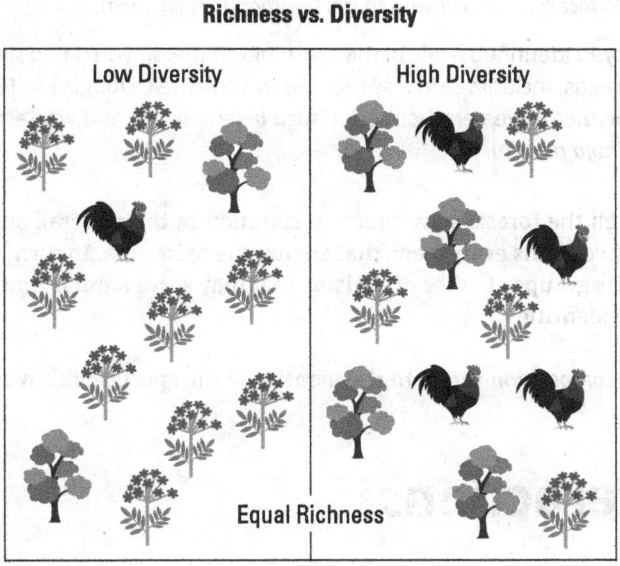

FIGURE 8-1: Species richness and species evenness.

Ecologists use their measurements of species richness and abundance to calculate values that represent the diversity of different communities. The most commonly used value for species diversity is the *Shannon index*, a statistical calculation that includes both species richness and relative abundance.

REMEMBER

The higher the value of the Shannon index, the higher the species diversity in a community.

Identifying species composition

The species that live in a community help define its structure. If you were to walk through two temperate rainforests, one in New Zealand and one in Washington state, USA, you would see many similarities. Both are dominated by tall evergreen trees. Ferns and mosses fill in the available spaces, even covering the tree trunks with a fuzzy green coat. But if you took a close look, you would see that these forests have many differences. Some of the evergreen trees in the New Zealand forest make fleshy structures that resemble berries, while the evergreens in Washington make woody cones. The bird songs in the two forests are different, too, and so are the wildflowers:

- If you spent some time identifying the trees in the New Zealand rainforest, you would discover that the trees with the berry-like cones belong to a family of evergreens called podocarps that evolved in the southern hemisphere. The trees you're seeing might include Rimu (*Dacrydium cupressinum*), totara (*Podocarpus totara*), and matai (*Prumnopitys taxifolia*).
- If you identified trees in the Washington forest, you'd find different evergreens, including Sitka spruce (*Picea sitchensis*), Douglas fir (*Pseudotsuga menziesii*), western hemlock (*Tsuga heterophylla*), and western red cedar (*Thuja plicata*).

Although the forests have much in common in their overall structure, each community contains evergreens that are unique to its area. In turn, each group of trees supports groups of other organisms that may serve similar functions but also have unique identities.

REMEMBER

Species composition refers to the identity of the species within a community.

Tracing Interactions

Species in communities interact directly and indirectly with other species:

- *Direct interactions* occur between two species and include things like competition, parasitism, and predation (see Chapter 7 for more examples). A snail grazing on algae in the intertidal zone is an example of a direct interaction.
- *Indirect interactions* occur when a direct interaction impacts additional species. For example, a seagull may pull out and eat a snail, then drop the empty shell, providing a potential home for a hermit crab. The predation by the seagull on the snail is a direct interaction, but it leads to an indirect interaction with the hermit crab.

Indirect interactions can have far-reaching impacts in a community. The strength of the impact can also be influenced by the particular species involved.

Meeting some major influencers

Some types of species have particularly large impacts on community structure:

- » ***Dominant species* are the most common or largest species in a community.** They may affect community structure by outcompeting other species for resources or by altering environmental conditions such as nutrient and water availability. A dominant tree species in a forest may capture most of the available light, creating shaded conditions that prevent understory plants from growing. In alpine environments, low-growing cushion-like plants trap heat and moisture that enable other species to grow near them.
- » ***Foundation species* create or maintain the physical or chemical environment that other organisms depend on.** Hard corals are animals that form protective outer skeletons of calcium carbonate. The bodies of hard corals form elevated reefs on the ocean floor and provide habitat for thousands of other species. Although coral reefs only form about one percent of the ocean floor, they support 25 percent of all marine species. Corals form symbiotic partnerships with photosynthetic algae that help the corals thrive in nutrient-poor water. Warming ocean temperatures cause corals to expel their algae, resulting in coral bleaching and death. Without the corals, the reef community collapses. Figure 8-2 compares the structure of Caribbean coral reef communities from the 1970s to today. Staghorn and elkhorn coral formed the foundation for a diverse community in the 1970s, but as they began to die, the community collapsed. Foundation organisms like coral that modify the environment in ways that affect which organisms live there are also called *ecosystem engineers*.

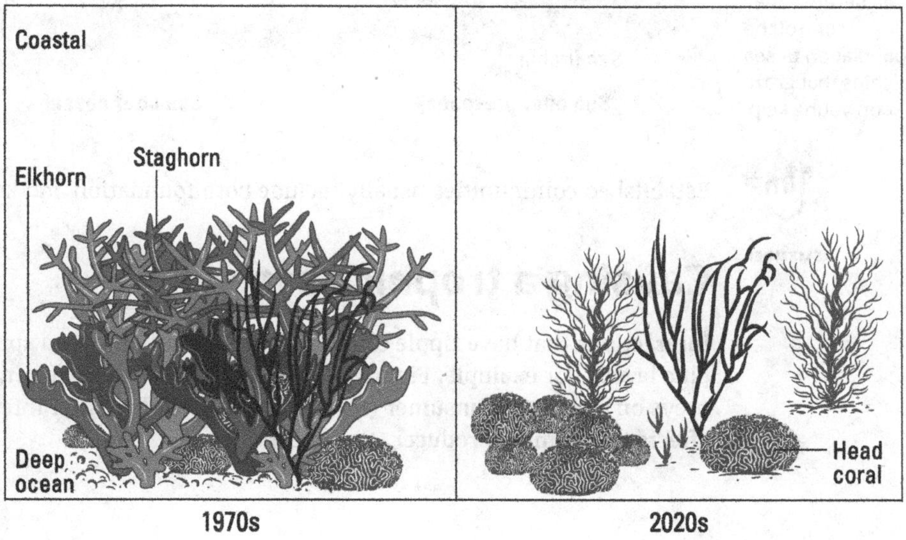

FIGURE 8-2: Foundation species. Staghorn and elkhorn coral are foundation species in Caribbean coral reef communities. If they are lost, the community collapses.

» ***Keystone species*** **keep a community in balance by controlling populations of other species.** Keystone species (see Figure 8-3) may not be large or abundant, but their loss leads to big changes and even the collapse of a community. Sea otters are keystone species in kelp forests because they prey on sea urchins, controlling their population growth and preventing them from overgrazing the young kelp. Without sea otters, sea urchin populations expand, and the kelp forest is destroyed. Increased predation by orcas on sea otters in the Aleutian Islands removed the keystone species, allowing sea urchins to destroy kelp forests and damage reefs made by coral-like red algae.

FIGURE 8-3: Keystone species. Sea otters are a keystone species in kelp forests because they control the population of sea urchins that graze on young kelp.

REMEMBER

Established communities usually include both foundation and keystone species.

Causing a trophic cascade

Interactions that have ripple effects within a community can spread between trophic levels. For example, Figure 8-3 shows how a secondary consumer (the otter) preys on a primary consumer (the sea urchin) and how that interaction indirectly affects the primary producer (the kelp).

![REMEMBER icon] REMEMBER

Trophic cascades occur when interactions between organisms at one trophic level indirectly affect species at different trophic levels.

Trophic cascades have the potential to reshape community structure. Figure 8-4 shows how control of a trophic cascade can come from an organism at the top of the trophic pyramid, as in the sea otter example, or it can come from the bottom (for more details on trophic pyramids, head to Chapter 11). In many cases, including the one in Figure 8-4, both types of control influence community structure.

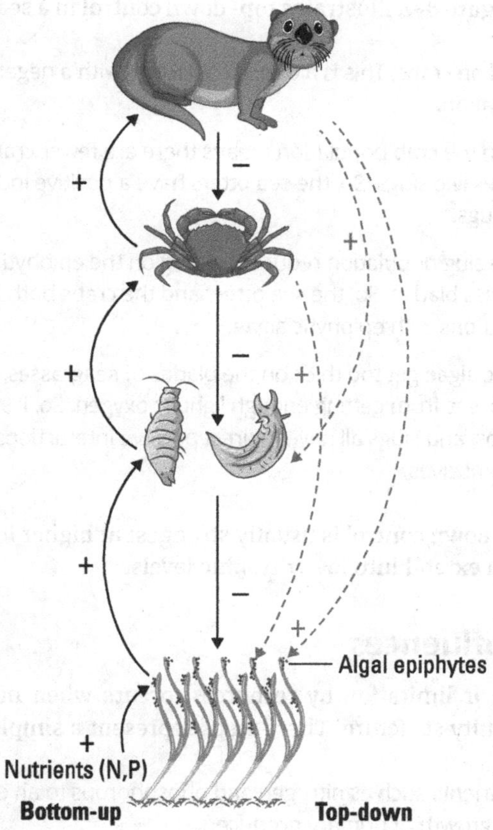

FIGURE 8-4: Trophic cascades. Solid lines represent direct indirections. Dashed lines represent indirect interactions.

Top-down control

Top-down control, or limitation by enemies, begins when one organism feeds on another. These steps represent a simple example:

1. Secondary consumers feed on primary consumers, limiting their population growth.

2. The reduction in primary consumers results in lower herbivory on primary producers.

3. Primary producer populations benefit from the reduction in herbivory.

TIP

One way to think about top-down control is with the phrase "the enemy of my enemy is my friend." Herbivores are the enemy of primary producers, so the predators that control herbivore populations act as friends to the primary producers.

The right side of Figure 8-4 illustrates top-down control in a seagrass community.

1. Sea otters feed on crabs. This is a direct interaction with a negative impact on the crab population.

2. The decrease in the crab population means there are fewer crabs to feed on small herbivores like slugs. So, the sea otters have a positive indirect interaction with the slugs.

3. Decreasing the slug population reduces grazing on the epiphytic algae that grow on seagrass blades. So, the sea otters and the crabs both have positive indirect interactions with epiphytic algae.

4. When epiphytic algae get too thick on the blades of seagrasses, they may prevent seagrasses from getting enough light or oxygen. So, it's possible that sea otters, crabs, and slugs all have indirect positive interactions with the seagrasses themselves.

REMEMBER

The impact of top-down control is usually strongest at higher levels in the trophic pyramid, but it can extend into lower trophic levels.

Bottom-up influences

Bottom-up control, or limitation by resources, occurs when nutrient availability influences community structure. These steps represent a simple example:

1. Addition of nutrients such as nitrogen and phosphorous to an ecosystem stimulates the growth of primary producers.

2. Increased biomass of primary producers provides more food for primary consumers.

3. Increased numbers of primary consumers lead to increases in secondary consumer populations.

The left side of Figure 8-4 illustrates bottom-up control in a seagrass community:

1. Increased nutrients such as nitrogen and phosphorous in the water boost the growth of epiphytic algae that grow on seagrass blades.
2. Increased biomass of epiphytic algae supports the growth of herbivores like slugs.
3. The slugs are prey for crabs, so the crab population increases.
4. The crabs are food for sea otters, so they support that population.

REMEMBER

The impact of bottom-up control is usually strongest at the lower trophic levels. However, when increased resources lead to increases in net primary production from photosynthesis, biomass can increase at all trophic levels, and food chains may become longer.

Examining human interactions in communities

Normally, if a person talks about their community, they're talking about the other people who live in their local area. Through the lens of ecology, however, a person's community consists of all the living things in their area: the plants, animals, fungi, and even the microbes. Every action a person takes has the potential to influence their community, and the combined impacts of people can cause large-scale changes:

- **Adding fertilizer to the environment can trigger bottom-up changes to ecosystems.** On a small scale, people often add fertilizer to their lawns or gardens to support the growth of desired plants. On a larger scale, agricultural runoff into rivers and lakes can trigger massive algal blooms, leading to eutrophication and death of aquatic species (for details on eutrophication, flip to Chapter 10).
- **People prey on a wide variety of species, which has the potential to trigger top-down impacts on communities.** Fishing and hunting directly impact the species that are caught for food. It also has indirect negative effects on other species that compete for the same food and indirect positive effects on species that are prey for the catch. People occasionally hunt a population or even an entire species to extinction, which can dramatically alter community structure. A good example of this is the killing of wolves in Yellowstone Park, which you can read about in the sidebar "Where the Wild Things Live."

» **Humans are the dominant species in their communities, and they shape the environment in ways that favor some species but not others.** Farming causes large-scale environmental changes that completely alter community structure. For example, less than four percent of the original tallgrass prairie community survives in the United States. The rest was destroyed as land was modified for farming or urban development. When humans modify land, they fragment natural habitats, making it harder for other species to migrate to find food or mates (Chapter 13), ultimately leading to population declines. On a global scale, human activities are driving global climate change, increasing temperatures, and changing weather patterns. These changes have the potential for massive indirect impacts on community structure as species either die or migrate to new areas.

REMEMBER

Changes in environmental conditions have strong consequences for community structure because they modify species richness and abundance in both terrestrial and aquatic communities.

IN THIS CHAPTER

» **Evaluating sources of disturbance**

» **Rebuilding communities through succession**

» **Influencing the outcome of succession**

Chapter 9
Responding to Change: Disturbance and Succession

A hot spot of volcanic activity burns through the Earth's crust and slowly releases magma into the Pacific Ocean. The magma cools into basal rock that builds up until it breaks through the ocean surface, forming a rocky island. The island is completely bare and without life for years, but eventually, small patches of green appear as lichens or plants begin to grow. As they grow, these early pioneers make changes to the environment that enable other plants to live there. Animals arrive and create further change, often helping to build soil and disperse seeds. Over long periods of time, what was once an inhospitable pile of rocks becomes a lush tropical island covered with thriving communities that include coastal shrublands, lowland dry forests, bogs, upland wet forests, and alpine communities.

This type of transition from bare rock to thriving ecosystems occurs today as new islands form in Hawaii and the Galapagos. This process of community change, called *succession,* also happens around the world when established communities are disturbed by things like weather, fire, and human activity. In this chapter, you will look at how and why succession occurs and explore some of the ways that ecologists study this process.

Sources of Disturbance

Communities change over time, sometimes gradually with slow environmental changes and other times dramatically in response to large-scale disturbances.

REMEMBER

A *disturbance* is anything that causes a sudden change in an ecosystem, usually by altering the environment and access to resources.

Disturbances can be caused by non-living forces or biological activity.

Forces of nature

You've probably witnessed or been impacted by at least one major abiotic disturbance in your life:

- **Severe weather** triggers events like flooding, drought, and high winds that have the potential to completely change the course of succession. Hurricanes cause erosion in coastal areas and may flood breeding centers for birds. High winds can damage trees, especially those that lack stable root systems, and blow flying animals into new areas. In 2005, when Hurricane Katrina struck New Orleans, wind speeds of 140 miles per hour knocked down trees and stripped leaves off remaining vegetation. In total, scientists estimate that Katrina killed or damaged 320 million trees in Louisiana and Mississippi. Wildlife that live in areas with frequent storms may have adaptations that help them survive and rebound; however, these defenses can fail in the face of intense storms. Environmental damage from severe weather is especially dangerous to species that are already at risk due to habitat loss.
- **Geological events** like earthquakes, volcanic eruptions, avalanches, and mudslides can remove dominant species and sometimes even all life from the areas they impact. (For a description of how volcanic eruptions can affect communities, see the "What Happens When a Volcano Blows Her Top?" sidebar.)
- **Fire** is one of the most common disturbances of natural ecosystems. It directly kills many organisms, especially those that can't escape. Some organisms have traits that give them fire resistance, such as thick bark on some trees. Seeds that are buried in soil may also survive. Organisms with traits that confer fire resistance may have an advantage as succession begins. Some pines seal their cones with resin or wax that melts away during the fire, so that they release the seeds after the fire. Because fire creates disturbance and also favors fire-resistant species, it can have a strong influence on ecosystems that experience frequent fires.

Biotic disturbances occur when organisms disrupt community structure:

- **Ecosystem engineers** like beavers, pocket gophers, and prairie dogs make changes to the physical environment (for more on ecosystem engineers, flip back to Chapter 8). Beavers build dams that slow streams, creating ponds and wetlands that provide important resources for fish and amphibians. Prairie dogs dig systems of tunnels called prairie dog towns that are home to dozens of species, at least one of which can't live anywhere else. Scientists estimate that a single pocket gopher moves about 18 pounds of soil every year, creating spaces for new plants to colonize, aerating the soil, and moving seeds around.
- **Herbivory** by large animals such as deer, elk, and wildebeest can cause dramatic changes to plant communities. If grazing on grasslands is too intense, perennial grasses may be killed, allowing other species to move into the community. In the American West, overgrazing by cattle has allowed the invasion of cheatgrass, a non-native annual. In some areas, cheatgrass is now the dominant grass. Cheatgrass dies and becomes highly flammable, increasing fire risk just before the start of fire season. Fires create more disturbance and opportunity for cheatgrass to take over. On the African savannah, grazing by wildebeest changes the plant community in ways that facilitate other grazers. Wildebeest graze on tall grass, making it easier for Thomson's gazelles to graze on short grass and other plants.

Human impacts

Humans create disturbances when they extract materials from ecosystems or convert them for human uses such as agriculture and urban development. Two common types of anthropogenic disturbances are logging and farming:

- **Logging** impacts vary depending upon the method. Clear-cutting, the practice of removing every tree from an area, causes drastic changes to the environment, increasing light and erosion and decreasing moisture and organic material. The removal of trees means loss of animal habitat and stream protection. Even if the site is replanted, a forest that grows from saplings that are the same age and same species won't reproduce the original forest ecosystem. In selective logging, people cut mature, high-value trees from the forest, but leave the rest. This practice opens up smaller disturbance areas to new species. In general, scientists find that all types of logging affect community structure but clear-cutting does the most damage to the environment and has the largest negative impact on species richness.
- **Agriculture** frequently begins with removing the original community from an area through cutting trees, slashing and burning plants, or cutting through

tough grass roots on a prairie. The original community is replaced with plantings of the desired crops. A single species usually dominates each field. These changes seem obvious, but what is less visible are the changes to the soil community, increases in erosion, and, in some countries, the addition of fertilizers, pesticides, and herbicides to the local ecosystem. These chemicals can contaminate groundwater and flow in runoff into nearby streams and rivers. When farmlands are abandoned, new communities develop in the old fields.

Responding to Disturbance

Disturbances cause environmental changes that challenge the stability of community structure. How a community responds depends upon the intensity of the disturbance and the ability of organisms to survive the change:

- ***Resistance* is the ability of a community to maintain its structure in the face of disturbance.** For example, hardwood forests dominated by post oaks are drought-resistant and have a reputation for growing in poor soils where other forests can't survive.
- ***Recovery* is the speed at which a community rebuilds its structure after a disturbance.** Lodgepole pine forests recover quickly after fires because the pines have adapted to this disturbance. Lodgepole pines protect their seeds by sealing some of their cones with resin. When a fire passes through the forest, the resin melts, and the cones release the seeds. The seeds sprout, beginning the process of recovery.

REMEMBER

Resilience is the ability of an ecosystem to maintain its structure and function after a disturbance. In other words, it's a measure of the system's ability to bounce back after disturbance through a combination of both resistance and recovery.

Sometimes a community can neither resist nor recover from a disturbance. If disturbances are too frequent or too intense, they may change the environment so dramatically that an entirely new community will form in place of the old.

Types of Succession

Communities establish themselves and recover from disturbance through the process of succession, during which species colonize an area and then are replaced by the next wave of species, as shown in Figure 9-1:

» *Primary succession* occurs on land that initially lacks organic soil or living things. Today, primary succession occurs on newly formed islands, like those in Hawaii or the Galapagos, or in areas after a catastrophic disturbance like a volcanic eruption. (See the "What Happens When a Volcano Blows Her Top" sidebar for a story about succession after a volcanic blast.)

» *Secondary succession* occurs on land that has some organic components in the soil, but where a disturbance has removed most of the plants. Secondary succession restores communities after disturbances such as fire, logging, and disturbance by large animals.

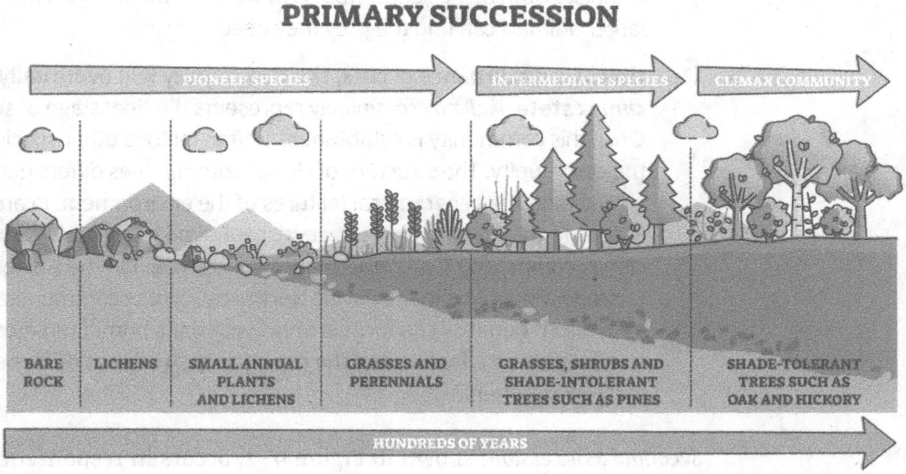

FIGURE 9-1: Primary succession.

VectorMine/Adobe Stock Photos

How to succeed

Primary succession introduces life into an area that has none, such as recently cooled lava. These environments may be little more than bare rock without soil. Organisms colonize and then change this environment, leading to a process of community change that occurs over hundreds of years, as shown in Figure 9-1:

1. ***Pioneer species* are the first to arrive via seeds or spores delivered by the wind, passing birds, or by the waves to a coastline.** Pioneers have characteristics that let them survive in this particularly harsh environment. Lichens make good pioneers because the fungal partner protects against water loss, and the alga provides food through photosynthesis (see Chapter 7 for more details on this type of symbiosis). Drought tolerance and shallow rooting systems allow some cacti to survive with very little soil.

2. **As pioneer species grow, they alter the environment.** They help trap moisture, create some shade, and when they die and decompose, they start the accumulation of soil.

3. **New species of plants and animals move into the environment, continuing the process of change.** Small annual plants can grow in the developing soil, and animals such as spiders and insects find the food and moisture they need.

4. **As conditions continue to change, new species move in, and old species leave the community.** Soil development allows larger plants like shrubs and small trees to eventually grow, but their shade can prevent some of the earlier colonists from being successful. As more small animals join the community, larger animals can find the prey they need.

5. **Unless a disturbance occurs, the community will eventually reach a *climax* state.** A *climax community* represents the final stage of succession. Once this community is established, it often inhibits other species from joining the community. The structure of climax communities differs depending on climate and the geographical features of the environment. In areas of the Blue Ridge Mountains with high elevation, cool climate, and abundant rainfall, the climax community is a northern hardwood forest. On the western slopes of the Cascade Mountains in the Pacific Northwest, climax communities dominated by western hemlock can form in areas with mild, humid climates. Both climax communities are forests, but the community structures differ because of the local environment.

Secondary succession, shown in Figure 9-2, occurs in response to disturbance. The community returns to an earlier point in succession, but not all the way back to the barren landscape. Soil, microbes, nutrients, and seeds are already present. Fire triggers secondary succession when it burns away most of the living things above the soil surface but leaves behind survivors like microbes and seeds that were protected by soil or fire-resistant pinecones. Secondary succession occurs frequently in ecosystems because disturbances happen often.

REMEMBER Succession drives vegetation change; disturbance turns the clock back.

REMEMBER The original community has a strong influence on secondary succession because of the soil structure and potential for survivors, called *biological legacies*, to regrow.

In both types of succession, the organisms present in the community may influence the success of other organisms:

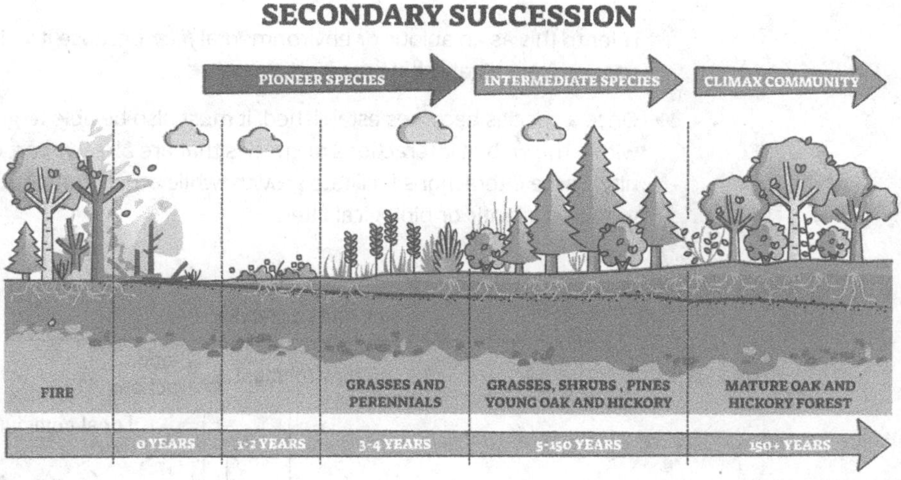

FIGURE 9-2: Secondary succession.

VectorMine/Adobe Stock Photos

- ***Facilitation* occurs when one species helps another species join the community.** When pioneer species begin the development of soil, they change conditions so that other species may grow. Some early colonizers form associations with bacteria that add nitrogen to the soil, which can make it possible for later colonizers to establish themselves.
- ***Inhibition* occurs when one species negatively affects the establishment of another.** The taller trees that are part of climax communities shade the environment, preventing many smaller species from growing. More direct inhibition occurs when one species eats another or if a prey species produces chemicals that prevent digestion in the predator.
- ***Tolerance* describes the situation of a species whose success is independent of other members of the community.** As early successional species grow and reproduce, other species establish based on their ability to survive the local environmental conditions. When the early species die out due to environmental or biological stress, tolerant species remain.

The influences of biotic interactions and the environment on community structure are incorporated into an explanation called *community assembly theory*, which is illustrated in Figure 9-3. After a disturbance opens up space in an area, the structure of new communities is influenced by several factors:

- Potential colonists from the region must disperse to the new community. Species that have better dispersal mechanisms are more likely to reach new areas.
- If dispersal brings a new organism to the site of the disturbance, it must be able to survive the environmental conditions there. Ecologists sometimes

refer to this as an abiotic or environmental *filter* because it will prevent some species from establishing themselves.

» Once a species becomes established, it must also be able to grow successfully within the web of interactions of species that are already part of the community. Some interactions facilitate growth, while others inhibit it. Ecologists refer to this as a biotic or biological filter.

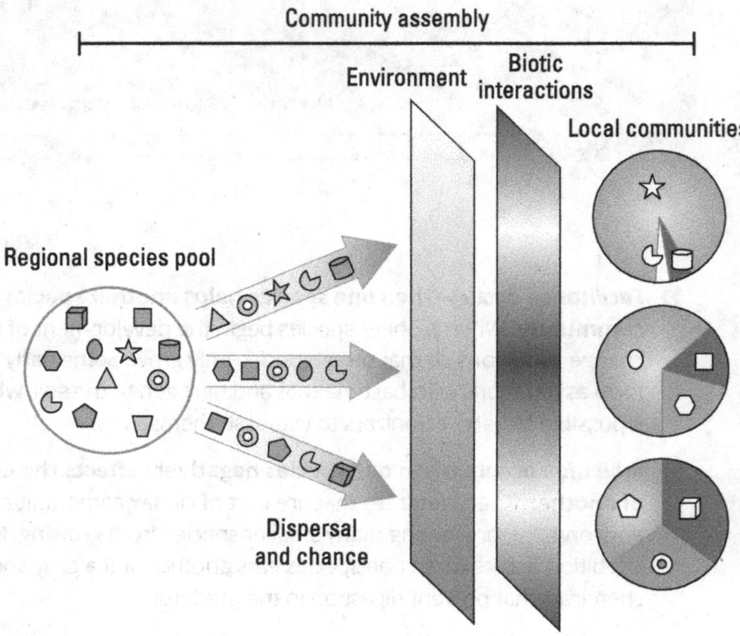

FIGURE 9-3: Community assembly theory.

REMEMBER

For a species to become established, it must be able to tolerate the environmental conditions. Once established, its success may depend upon its interactions with other species.

REMEMBER

Random chance, timing, and small differences in geography and climate may combine to create variations in local communities even within the same region.

It's the journey, not the destination

The view of succession proceeding toward a climax community traces back to the roots of ecology. In the 1890s, University of Chicago botanist Henry Chandler Cowles took a walk away from the shore of Lake Michigan, over the smaller sandy dunes and up a hill until he reached the top of the tallest dunes. As he walked, he noticed that the vegetation changed steadily as he got further from the lake.

He thought that the area closest to the lake represented the community that began on bare sand. As he walked further away, he reasoned that the communities he was seeing had been developing longer. In other words, as he moved physically in space, he was seeing the effects of passing time. Ecologists today call this the *space for time substitution.*

Today, you can walk the Dune Succession Trail in Indiana Dunes National Park and see the landscape that inspired the idea of succession:

» **Few species grow on the sandy dunes closest to the lake, which are called fore-dunes.** The most important dune-building species is marram grass, also called beachgrass. Marram grass grows by underground stems called rhizomes that extend out, stabilizing the dunes. These rhizomes also contribute to its ability to survive even when it gets covered by sand. A few Eastern cottonwoods grow from sheltered spots among the dunes. Once these trees get established, they survive sand deposits by turning buried stems into new roots. As bits of the plants die, they decompose and add organic matter to the sand, beginning the process of developing soil.

» **New plants appear in the more sheltered open dunes.** Little bluestem grass, which doesn't survive being buried in sand, can grow here. In between thick clumps of grass, various wildflowers appear, including the endangered dune thistle. Small shrubs begin to appear. The plants provide shade, lowering the temperature of the sandy soil that continues to develop with the addition of organic material.

» **As you move further upland, you spot jackpine trees growing in the nutrient-poor sandy soil.** Bearberry and juniper grow below the pines, and squirrels and birds make their homes here.

» **As you continue to climb, oak trees begin to invade under the pine trees, and other hardwood trees appear as the climax community takes form.** As you walk from the pine forest toward the oaks, you notice more plant debris scattered on the sandy soil, which changes in appearance and texture until it appears rich and dark. The hardwood forest contains red and black oak, as well as basswood and witch hazel trees.

This particular pattern of succession is unique to the Indiana Dunes. Still, similar progressions have been found in environments all over the world, leading to climax communities specific to the climate and geology of the area.

A recipe for species richness

Although many climax communities exist in nature, disturbance and secondary succession are happening everywhere. Communities don't march inevitably toward climax; they get reset by disturbance, then re-establish, or sometimes take a different fork in the path toward a new community structure. Succession is a constant ecological process, but the future of a community isn't known.

REMEMBER

Cyclical or random disturbances keep communities from reaching their climax state.

Disturbance may also play a role in determining species richness within communities. Figure 9-4 illustrates this idea:

- **If disturbances are small or infrequent, as shown on the left side of the graph, species richness will be low.** Communities will progress toward their climax state, and the dominant species will inhibit other species.
- **If disturbances are very large or frequent, as shown on the right side of the graph, species richness will again be low.** The only members of the community are likely to be pioneer species that can tolerate these stressful conditions.
- **The greatest levels of species richness would result from disturbances that were neither too frequent nor too large, which is shown in the middle of the graph.** Small or infrequent disturbances open up opportunities for new species to enter without disrupting the entire community.

REMEMBER

The *intermediate disturbance hypothesis* proposes that intermediate levels of disturbance lead to the greatest species richness in communities. Some, but not all, experiments provide support for this model, so ecologists are still testing this idea today.

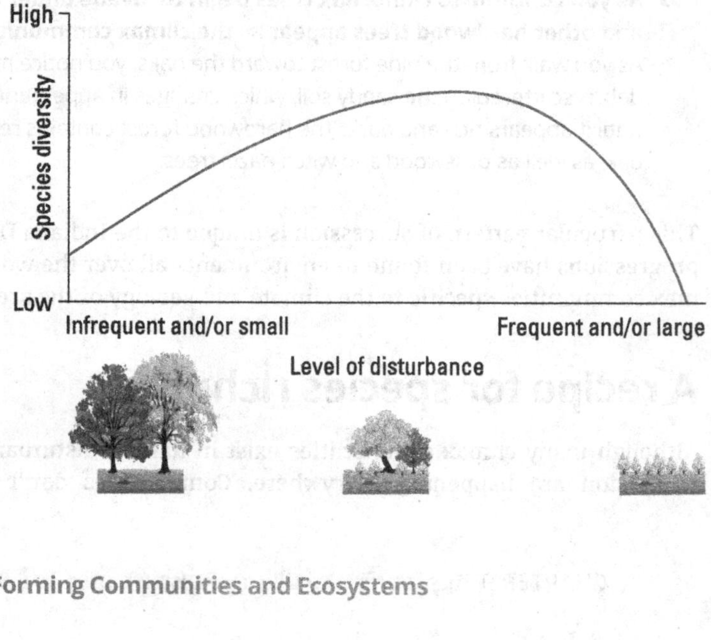

FIGURE 9-4: The intermediate disturbance hypothesis.

Studying Succession

Understanding the interplay between disturbance and succession may be more important to humans today than ever before. We depend upon ecosystems around us for many essential resources, including food, materials, and clean water. Human activities create disturbances that alter the communities and ecosystems around us. When communities change, we may lose access to certain resources and even destabilize systems beyond their ability to recover. The more we understand disturbance and succession, the better we will be able to manage our own impacts on ecosystems so that they will continue to support human life and the lives of the species around us.

Watching succession in real time

Ecologists may make direct measures of ecological succession by monitoring an area over time as it goes through succession. Experimental areas may occur naturally or be created experimentally:

» After a natural disturbance such as a volcanic eruption or hurricane, ecologists observe the affected area as it changes over time. Ecologists have been studying succession on Mount St. Helens ever since the volcano erupted over 45 years ago. After hurricanes Katrina and Rita hit Louisiana in 2005, ecologists observed several different marsh communities in the area over a period of two years. They compared their observations to pre-hurricane data in order to understand how the hurricanes impacted community structure.

» Another way for ecologists to study the impact of disturbance is to create one for themselves, often by removing organisms from a study area. Ecologist Terence Farrell studied succession in the rocky intertidal zone along the Oregon coast and found that acorn barnacles facilitated colonization by large algae:

- First, Farrell scraped rocks clean of organisms and watched them over time. The first species to move into his experimental patches were small brown barnacles, followed by acorn barnacles, and then several species of algae. The little brown barnacles were the pioneer species, but it wasn't clear what factors were influencing the later colonists. Farrell wanted to know if the barnacles were facilitating or inhibiting the colonization by algae in some way.
- Farrell set up additional plots, some where he removed just the small brown barnacle, some where he removed the acorn barnacles, and others where he removed both barnacles. He also set up control plots where all organisms were removed.

- Farrell observed his plots for two years to see how his treatments affected the ability of algae to colonize the rocks.
- The plots where just small brown barnacles were removed were almost the same as the control plots, showing that the small brown barnacles didn't impact algal colonization.
- In plots where the acorn barnacles were removed, including the plots where both barnacles were removed, the algae couldn't colonize the rocks. These experiments showed that the acorn barnacles facilitated colonization by the seaweed.

Taking a walk into the past

Ecologists study succession indirectly by monitoring areas along a gradient that represents time since succession began. For example, scientists monitored plots on eight different lava flows on the slopes of Mauna Loa volcano in Hawaii. Each study area was very similar in environmental conditions except for the time since the lava flow disturbance, which ranged from eight to 9,000 years before the study was conducted. Although the scientists were studying all the plots simultaneously, the community in each plot represented a different time point in the succession journey.

REMEMBER

A *chronosequence* is a series of ecological sites that are extremely similar but represent different ages.

Using the chronosequence method, the ecologists were able to identify pioneer species and describe four distinct successional phases. They also noted that diversity in the communities increased for about 300 years, then remained relatively steady. (Diversity was measured with the Shannon Index, which is discussed in Chapter 8.)

WHAT HAPPENS WHEN A VOLCANO BLOWS HER TOP?

On May 18, 1980, an earthquake in Washington state triggered a series of violent events around the volcano known as Mount St. Helens. The largest landslide in history ripped down the volcano's north side, depositing debris over 24 square miles. This was accompanied by a lateral blast of rock, ash, and hot gas to the north, causing devastation over an additional 150 square miles. The rivers that flowed down from the volcano became lahars, flows of hot debris and water that resembled liquid concrete. Overlapping waves of hot gas, magma, and volcanic rock, called pyroclastic flows, poured out over the north face. A vertical plume of ash spewed upward, catching the air currents and drifting eastward for more than 125 miles. When the violence finally subsided, Mount St. Helens was completely transformed. The beautiful snow-capped mountain that was home to many plant and animal communities and overlooked the clean jewel of Spirit Lake, became a seemingly barren gray hulk, covered with rock and ash and topped with a sunken crater. Fifty-seven people were killed, and 42 square miles of trees died, many scattered like toothpicks across the volcano's surface. Thousands of animals were killed in the blast.

Ecologists were among the first people to return to the slopes of the volcano, arriving just ten days after the blast. From the air, the volcano looked like a gray, lifeless hulk, and the scientists thought that all life had been eliminated near the blast. Their very first footsteps proved them wrong when they displaced the ash to reveal tiny green shoots of fireweed, a well-known pioneer species that sends up tall shoots of bright flowers. One of fireweed's traits that makes it a good pioneer is that it can regrow rapidly from small root fragments. The scientists realized with surprise that some roots of fireweed plants survived the blast and were already beginning the process of rebuilding the community.

Fireweed was the first survivor to be discovered, but the ecologists soon found more, including shoots of thistle and pearly everlasting, and beetles scurrying among the debris. Fresh dark dirt pushed up through the ash, signaling the survival of pocket gophers whose underground burrows might have protected them. Likewise, ice and snow on lakes and rivers had protected some aquatic life, and the scientists saw green algae beginning to spread in the river. Western toads flourished when they emerged from underground hibernation into a community that lacked their normal predators. Tracks of mice and elk in the ash suggested animals were already moving in from the surrounding area. All of these events could be predicted by an understanding of secondary succession, but no one expected it to begin this fast.

On the north side of the crater, the destruction was more thorough. An area of six square miles, called the pumice plain, was reduced to bare rock and completely lifeless. Although the devastation was staggering, ecologists recognized that nature had presented them with a unique opportunity to study primary succession in action. Researchers planned studies, marked their plots, and returned at regular intervals to make observations.

In June 1982, almost two years after the blast, ecologists spotted a spray of purple flowers. They'd found their first pioneer plant, a prairie lupine. Lupines dramatically improve

(continued)

(continued)

soil quality by adding both nitrogen and phosphorous. They get nitrogen from their associations with nitrogen-fixing soil bacteria (see Figure 9-6) and can use their roots to access phosphorous from rocks. Lupines also provide shade and increase moisture in the local area. The lupines were the signal that primary succession was well underway.

It's been more than 45 years since the eruption, and the communities on Mount St. Helens continue to change. Plants on the pumice plain now include willows, mosses, wildflowers, grasses, huckleberries, conifers, and even maples. The animals include insects, birds, and pocket gophers, plus herds of elk that wander through to graze. Research continues on a wide variety of topics linked to succession, including soil development, animal tracking, forest growth, and feedback between biotic and abiotic factors. From the research, it's clear that succession doesn't happen smoothly from pioneer species to climax community.

The process is uneven and bumpy as new disturbances create opportunities for change. The abiotic factors in the many local environments on the volcano influence local communities, creating a patchwork of succession and diversity. The changes in the abiotic and biotic factors on the volcano were so dramatic that ecologists say the ecosystems there will never be the same as before the eruption. They continue to study the changes, building our understanding of the factors that influence succession so that we can apply that knowledge to our changing world.

Influencing Succession

Some organisms have particularly strong impacts on the way that community structure changes during succession. This may come from a characteristic that causes a species to facilitate or inhibit other organisms. Some animals have behaviors that affect the dispersal or growth of other species. Human beings sometimes deliberately influence succession, either in an attempt to restore a disrupted ecosystem or to keep a community in a desired phase of succession.

Early vs. late arrivals

For an organism to become part of a community, it needs to disperse to the location, grow in the environment, and then survive interactions with other organisms. Some species have characteristics that give them a head start on succession, while others are better at long-term survival:

- ***Pioneer species*** **typically disperse over distances and can grow rapidly in stressful environments. This enables them to be the first to arrive after**

a disturbance and begin the process of succession. Their growth often modifies the environment in ways that facilitate colonization by later species. Fireweed is an effective pioneer species due to its small, wind-dispersed seeds and its ability to grow in disturbed areas. It grows rapidly, spreading by underground rhizomes that stabilize the soil. These rhizomes may survive in the soil for long periods of time. If a disturbance clears the land so that lots of light becomes available, the rhizomes begin growing, allowing fireweed to rapidly dominate the area.

» ***Climax species*** **form a stable community after long periods of uninterrupted succession. These species may be larger and grow more slowly than pioneer species. Their growth often inhibits colonization by other species.** The yellow carabeen is a large, shade-tolerant tree that dominates late succession subtropical forests along the east coast of Australia. Its seeds germinate in shady, moist environments that would be found under shrubs and small trees. The seedlings grow slowly in the understory for long periods of time. If a tree falls and resources become available, seedlings will grow faster and take over the space. These trees can reach 180 feet in height and live for 800 years. They help create a shaded forest environment that prevents the establishment of species that aren't shade-tolerant.

Table 9-1 compares some of the common characteristics of plants that colonize communities early and late in the process of succession.

TABLE 9-1 Comparing early and late successional plants

Characteristic	Early	Late
Seeds	Many	Few
Seed size	Small	Large
Dispersal mechanisms	Wind, carried by animals	Gravity, eaten by animals
Seed viability	Long periods, remain dormant in soil	Short
Growth rate	Rapid	Slow
Size at maturity	Small	Large
Shade tolerance	Low	High

Animal helpers

The first studies of succession focused on transitions in the plant community, but animals are also part of the changing community structure. All through the successional process, animals may become established in the community once it provides the resources they need to survive. Figure 9-5 gives examples of some of the animals that might be found during succession toward an oak-hickory forest in North Carolina. Smaller or roving species are more common in the early phases, when plants are smaller. Once prey species are established, their predators may join during mid-successional phases. Larger species or those that need mature trees enter the community later.

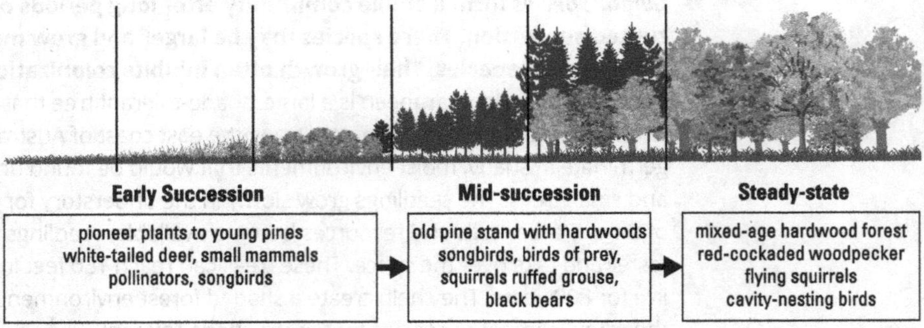

FIGURE 9-5: Animals that are commonly found during phases of succession in an oak-hickory forest.

Animals impact the process of succession in many ways:

- Small mammals spread seeds and are food for predators.
- Grazing animals like deer and elk limit some plants through grazing pressure while also depositing nutrient-rich feces that may contain seeds from neighboring areas and spores of mycorrhizal fungi. These fungi form associations with plant roots, helping plants grow better by increasing their access to water and minerals (for details on mycorrhizae, flip back to Chapter 7).
- Tunneling animals like pocket gophers and prairie dogs mix soil layers, distributing nutrients and oxygen throughout the soil. As they dig, they may bring plant seeds or rhizomes closer to the surface where they can grow. As they travel around the ecosystem, they may distribute seeds and spores of mycorrhizal fungi.
- Birds drop feces into the environment, providing additional nutrients for plant growth. They also help colonization by bringing seeds and spores from neighboring areas.

Nitrogen-fixers

Some pioneer species have a superpower that makes them especially good at colonizing nutrient-poor environments. They form associations with soil bacteria that have the ability to grab nitrogen out of the air and turn it into forms that the plants can use. Plants need nitrogen for their proteins and DNA, but nitrogen is often scarce in soil, especially after a disturbance. A lack of sufficient nitrogen will limit plant growth. (To find out more about the importance of nitrogen in ecosystems, head to Chapter 11.)

REMEMBER

Nitrogen-fixing bacteria capture nitrogen gas (N_2) from the atmosphere and convert it into ammonia (NH_3).

Nitrogen-fixing bacteria produce ammonia, use some as their own source of nitrogen, and then release excess ammonia into the soil, where it becomes available to plants. The bacteria live freely in the soil, and some of them can form symbiotic associations with plants (for details on symbiosis, check out Chapter 7).

Plants in the bean family, also called leguminous plants, form intimate associations with nitrogen-fixing bacteria, as shown in Figure 9-6, and described here:

- The roots of leguminous plants like lupines attract the bacteria, which attach to the root hairs and grow an infection thread that allows them to enter the root cells.
- The bacteria multiply inside the root cells, causing them to swell into visible *nodules*.
- Inside the nodules, the bacteria continue to fix nitrogen, sending ammonia directly to the plant.
- The plant makes sugars via photosynthesis, sharing some with the bacteria.
- Essentially, the plants have built-in fertilizer factories, and the bacteria have a steady income in the form of food.

In addition to leguminous plants, some trees in the birch family also form symbiosis with nitrogen-fixing bacteria. Red alder is common in disturbed areas in the Pacific Northwest because the symbiosis helps it survive where other species can't.

REMEMBER

Pioneer species that form associations with nitrogen-fixing bacteria can grow in nutrient-poor soils because they have a steady supply of nitrogen. When these nitrogen-rich plants drop leaves or die, their decomposition adds nitrogen to the developing soil, facilitating the growth of other species.

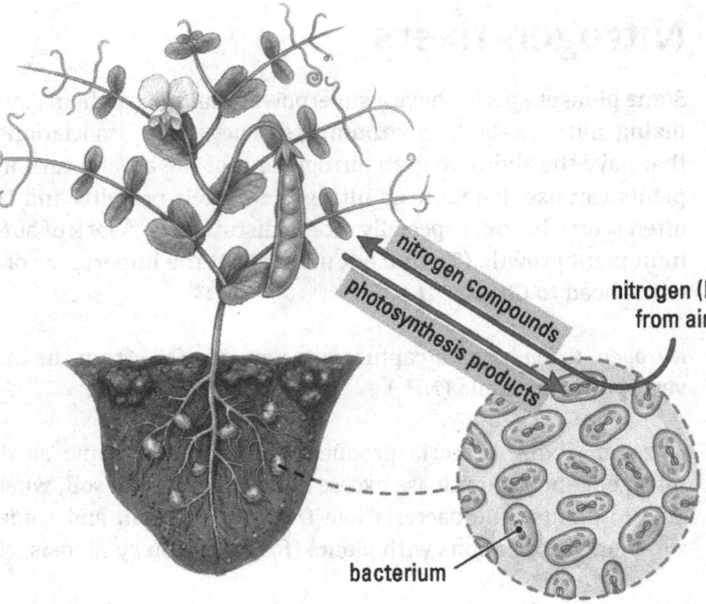

FIGURE 9-6: Nitrogen fixation in leguminous plants.

Invasive species

As humans and other species move around the globe, they sometimes bring other species with them. People may accidentally transport species like marine species on boat hulls. Or they may deliberately introduce a species as a garden ornamental or for the pet trade. If these species move into new ecosystems, they may disrupt community structure.

REMEMBER

Invasive species are non-native species that have the potential to cause harm in the ecosystem to which they are introduced.

Disturbance may provide an opportunity for an invasive species to establish itself. Once established, invasives have the potential to alter the process of succession:

- If an invasive species has traits that make it a good pioneer, it may get a head start on native species and inhibit them from getting established.
- Invasive species may change the environment in different ways than native species, so that facilitation of later species is lost.
- Invasive species may persist longer in the community than native species, inhibiting colonization and reducing species richness.

REMEMBER

Invasive species that are good pioneers or able to persist longer in an ecosystem have the potential for the greatest impact on the succession of resident communities.

REMEMBER

When community structure is altered by invasive species, essential ecosystem services may be lost to humans and other species.

Human management

People have a long history of using their understanding of disturbance and succession to manipulate ecosystems:

- Many indigenous people use controlled fire to introduce disturbance into small areas, causing plant communities to revert to an earlier stage of succession that supports the growth of important food plants.
- Many indigenous people follow guiding principles that limit how much they harvest from ecosystems, which has the effect of keeping disturbances to a minimum.
- Forest managers use lessons from successional studies to improve the health of forests. One example is the practice of leaving snags and fallen logs during logging because species that live within this wood will help regrow the community.
- In response to increasing wildfires, forest managers in the western United States have begun to use controlled burns to clear the understory from forests. Creating gaps between the trees helps prevent the rapid spread of fire across the treetops.
- People who work in ecological restoration are using their knowledge of local species and succession patterns to support the recovery of disturbed ecosystems. (For more on ecological restoration, head to Chapter 14.)

IN THIS CHAPTER

» Following energy through food chains

» Transforming energy from one type to another

» Depending on primary productivity

Chapter 10 Energy Flow through Ecosystems

Living things engage with each other and their environment in a complex web of interactions. Organisms seek the energy and nutrients they need to survive: some organisms can get what they need from the components of their environment that were never alive, while others take what they need by consuming the living or the dead. Each interacting community in a particular area forms an ecological system that functions as a unit. In this chapter, you will explore how organisms obtain and distribute energy through these ecological systems.

Examining Ecosystems

Organisms interact with each other and their environment to get what they need to survive. An evergreen tree captures energy from the sun and absorbs water from the soil with the help of a fungal partner around its roots. A red-crested bird makes a home in a dead tree and eats ants that live in the decaying wood. These organisms are part of a larger forest that is home to many different plants, animals, and microbes. All of the interacting parts of this forest, both living and nonliving, form an ecosystem.

REMEMBER

An *ecosystem* is an area of the Earth that includes all the living (*biotic*) and nonliving (*abiotic*) components of that area.

Scientists define the border of the ecosystem they want to study so that ecosystems can be as small as a group of microbes living in a hot spring or as large as the open ocean. Earth is unique among the known planets because it's covered with a huge ecosystem, called the *biosphere*, that includes all the environments on Earth plus all the living things that live near and on the surface. The unique nature of each ecosystem is shaped by abiotic factors such as climate and geography (for details, see Chapter 2), and defined by the types of organisms that live there. A major function of all ecosystems is the movement of energy through the system.

REMEMBER

Ecosystem ecology is the study of the movement of matter and energy through ecosystems.

Defining energy roles

Because the movement of energy is so important to ecosystems, ecologists often define groups of organisms by the role they play in the movement of energy (shown in Figure 10-1).

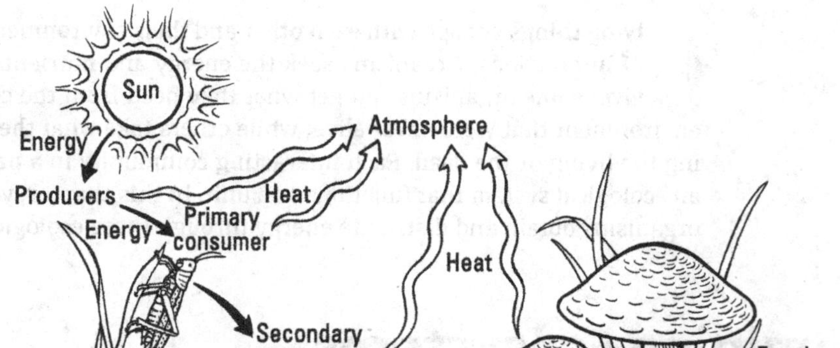

FIGURE 10-1: Energy flow in ecosystems is shown through a food chain.

» ***Producers* capture energy from abiotic sources and store it as chemical energy in food.** Scientists also refer to producers as *autotrophs* because they make their own food (*auto*=self, *troph*=feed).

- *Photosynthetic* producers, like plants, capture light energy from the sun, using it to make food from carbon dioxide and water. Ultimately, this food provides energy for all of the organisms in the ecosystem. In addition to plants, photosynthetic producers include organisms like seaweed, phytoplankton, and green bacteria.
- *Chemosynthetic* producers capture energy from abiotic chemical sources. These organisms are essential to life in ecosystems that lack light, such as the ocean floor and lightless caves. You may have heard about ocean communities that form around hydrothermal vents at the bottom of the ocean (shown in Figure 10-2). These communities could not exist without the chemosynthetic bacteria that use energy from the chemicals coming out of the vents to make food from carbon dioxide and water. Many of the animals around the vents, like the tube worms in Figure 10-2, provide a home for the chemosynthetic bacteria inside their own bodies. The tube worms capture chemicals from the water and pass them to chambers containing the bacteria. The bacteria make food for themselves and the worms. This close association between bacteria and worms is an example of symbiosis (flip back to Chapter 7 for more on symbiosis).

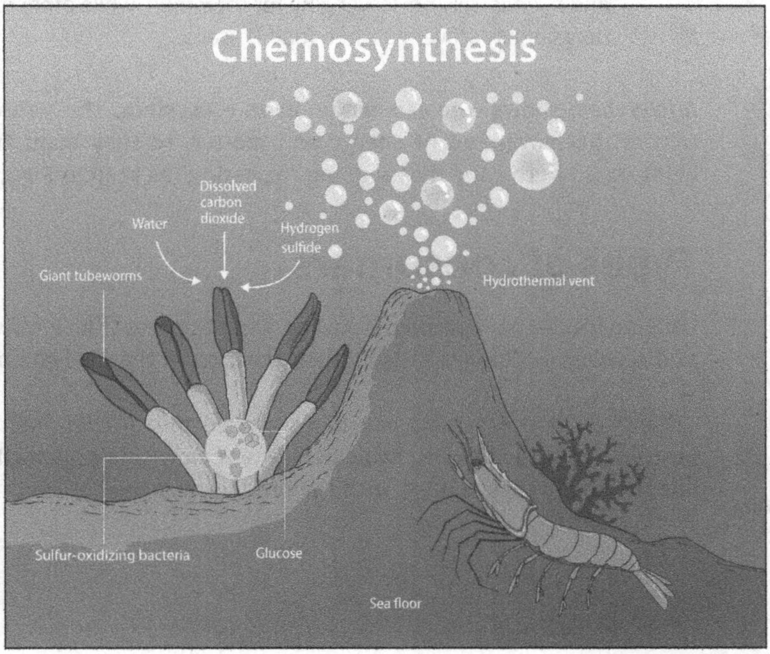

FIGURE 10-2: Chemosynthesis at a hydrothermal vent.

zombiu26/Adobe Stock Photos

- ***Consumers* get energy by eating other organisms.** Scientists also refer to consumers as *heterotrophs* because they must eat others (*hetero*=other).
- ***Decomposers, detritivores,* and *scavengers* are heterotrophs that get energy by eating the dead.** They break down the molecules in the remains of living things, taking energy and matter for themselves, while also releasing matter back into the environment in forms that other organisms can use (see Chapter 11 for more details on matter recycling). The main differences between these types of organisms are what they eat and how they feed.
  - **Decomposers are bacteria and fungi that release enzymes onto dead matter, digesting the food outside of their own cells, then absorbing the resulting small molecules.** These organisms are nature's ultimate recyclers because they produce a wide variety of enzymes capable of breaking down the toughest molecules made by living things.
  - **Detritivores are small animals that feed on bits of dead organic material, called *detritus*, that's found in soil and aquatic sediments.** Unlike decomposers, detritivores ingest, or bite, their food and then digest it inside their bodies. If you like crime shows, you may have seen scientists figuring out the time of death on a human cadaver based on the life cycle of flies or beetles found on the body. These insects, along with worms, snails, crabs, and many other small animals, are examples of detritivores.
  - **Scavengers are larger animals that feed on carcasses.** Like detritivores, they ingest their food and digest it internally. Some animals, like vultures and hyenas, rely on scavenging for their food, while other animals may occasionally scavenge when food is scarce.

REMEMBER

A *food chain* shows who eats whom in an ecosystem. The arrows show the direction of the movement of energy and matter, so they point from the organism being eaten to the one that is doing the eating (as seen in Figure 10-1).

Types of consumers

Consumers can be separated based on what they eat (like herbivores vs. predators as discussed in Chapter 7) or by their position in the food chain.

REMEMBER

Organisms in ecosystems can be organized into *trophic levels* that reflect their positions in a food chain. Ecologists often show this organization as a pyramid, like the one in Figure 10-3, with the producers at the bottom.

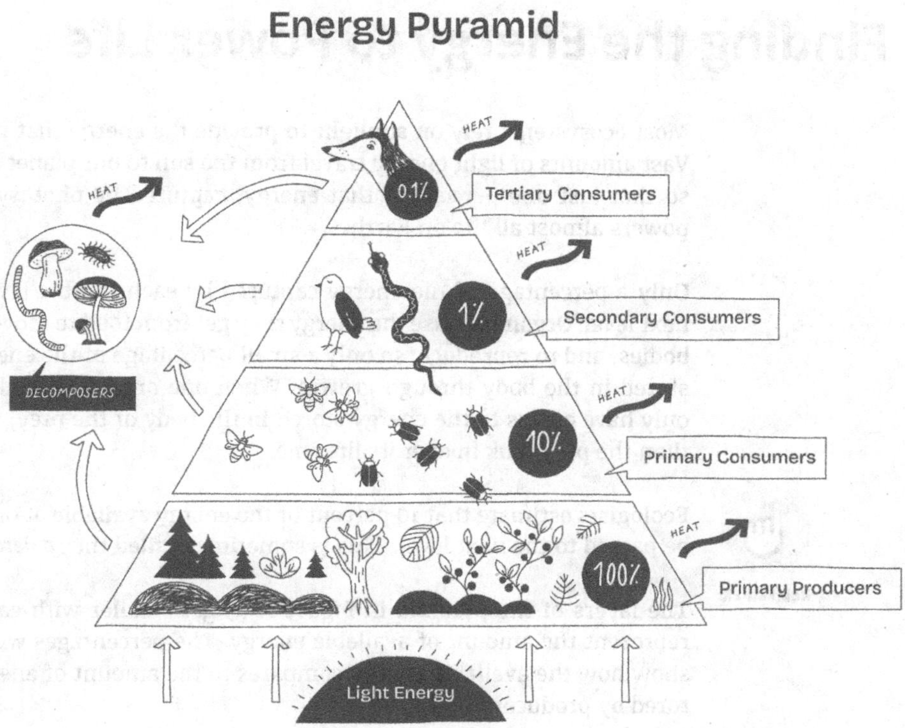

FIGURE 10-3: A trophic (energy) pyramid.

VectorMine/Adobe Stock Photos

Consumers are grouped in trophic levels that reflect their role in energy transfers in the ecosystem:

- ***Primary consumers* eat producers.** In the example shown previously in Figure 10-1, grasshoppers are a primary consumer because they eat grass (producers).
- ***Secondary consumers* eat primary consumers.** In Figure 10-1, a toad is a secondary consumer that eats grasshoppers.
- ***Tertiary consumers* eat secondary consumers.** A snake that eats a toad would be an example of a tertiary consumer.

Decomposers eat the dead from all the trophic levels, so they are usually shown off to the side of an energy pyramid, or sometimes they aren't included in diagrams of trophic pyramids.

Finding the Energy to Power Life

Most ecosystems rely on sunlight to provide the energy that makes life possible. Vast amounts of light energy travel from the sun to our planet every day, so much so that just one percent of that energy, captured by photosynthetic producers, powers almost all life on Earth.

Only a percentage of the energy captured by each trophic level is passed to the next level. Organisms use the energy they get from food to move, to maintain their bodies, and to reproduce, so only a small percentage of the energy they capture is stored in the body through growth. When one organism feeds on another, they only have access to the energy stored in the body of the prey, which is much less than the prey took in over its lifetime.

REMEMBER

Ecologists estimate that 10 percent of the energy available at one trophic level can be passed to the next level. This is sometimes called the *10 percent rule*.

The layers of the pyramid in Figure 10-3 get smaller with each trophic level to represent the amount of available energy. The percentages written on each level show how the available energy compares to the amount of energy originally captured by producers:

- Producers have access to 100 percent of the energy they capture from the sun.
- Primary consumers have access to 10 percent of the energy captured by producers.
- Secondary consumers have access to 1 percent of the energy captured by producers.
- Tertiary consumers have access to 0.1 percent of the energy captured by producers.

REMEMBER

Most ecologists think that the reduction in available energy at each trophic level is part of the reason that many ecosystems only support four to five trophic levels. The availability of energy at each trophic level also has implications for human populations, which you can read about in the sidebar "Diet for a Small Planet."

DIET FOR A SMALL PLANET

As the human population continues to grow, we need to answer the challenge of how to feed everyone. The World Resources Institute predicts that we'll need to feed 10 billion people by 2050, while also reducing our greenhouse gas emissions. This means we'll have to produce 56 percent more food without converting more land to agriculture. Trying to meet this goal is an enormous challenge, but one of the solutions they propose is something that people have been talking about since at least 1971, when Frances Moore Lappé published her book *Diet for a Small Planet*.

In her book, Lappé points out the reality represented by a trophic (energy) pyramid: eating higher on the trophic pyramid is not an efficient way to get energy. Because only 10 percent of available energy passes from one level to the next, eating meat is much less energy efficient than eating plants.

National Geographic says that for every 100 calories of grain we feed to animals, we can produce 40 new calories in milk, 22 calories in eggs, 12 calories in chicken, 10 calories in pork, or 3 calories in beef. One way to make sure we have enough calories to feed more people is to encourage people to shift to plant-based diets. If we consume more foods from the lowest trophic level, more energy will be available to us, and less will be lost to the inefficiency of energy transfer through trophic levels.

An ecologist named E.P. Odum created a diagram like the one in Figure 10-4 that helps visualize why so little energy is available from one trophic level to the next:

1. ***Ingestion* transfers food into the body when a consumer takes a bite of food.** Food contains stored chemical energy, so this brings energy into the organism.
2. ***Assimilation* occurs when the digestive system breaks down large food molecules and absorbs smaller molecules into the body.** When food enters an organism, it's frequently contained in a compartment. The food molecules don't actually become available for use until they're broken down and moved into cells. You start digesting your food as soon as it hits your mouth, and digestion continues as your food passes through your stomach and small intestine. Assimilation doesn't occur until cells of your small intestine move the broken-down food molecules into your blood, where they can circulate to the cells of your body.
3. ***Excretion* is the release of waste products into the environment.** Some waste products contain stored energy. For example, many animals can't digest plant fibers like cellulose. When fiber is released to the environment in feces, its energy becomes unavailable to the next trophic level, but it can still provide energy to decomposers that can digest cellulose.

4. ***Cellular respiration* transfers energy from food so that organisms can use it to do work like movement, repair, and transport of molecules.** Energy that's transferred to do cellular work ultimately ends up as heat energy in the environment, which is unavailable to the next trophic level.

5. ***Growth* occurs when food is used to build new molecules and structures.** Growth stores energy from food in a way that is accessible to the next trophic level, so ecologists sometimes refer to this as *secondary production*. Only *primary producers*, like plants, algae, and chemosynthetic bacteria, can actually make food from environmental sources.

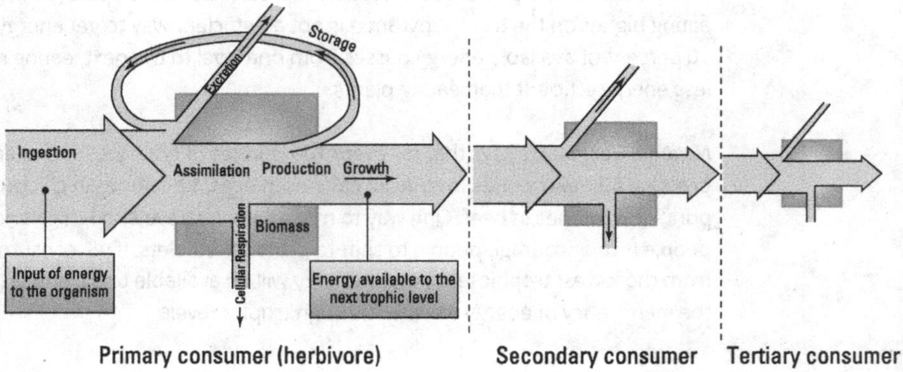

FIGURE 10-4: Universal model of ecological energy flow.

Following the rules

Scientists have studied energy in many different systems to figure out how energy transfers work and how to make them more efficient in human designs. Their observations led them to conclude that energy always behaves according to certain universal laws that scientists call the laws of thermodynamics. The first two laws of thermodynamics can help us understand how energy moves through ecosystems:

» **The *first law of thermodynamics* states that energy can't be created or destroyed.** This law tells us that living things can't make energy; they can only move it from one place to another.

» **The *second law of thermodynamics* states that the total disorder (entropy) of a system increases over time.** In terms of ecology and living systems, this law tells us that with every energy transfer, some energy is transferred to heat. In Figure 10-3, every level of the trophic pyramid represents the transfer of energy to heat with curved black arrows. Heat energy can't be used for cellular work, so the transfer to heat represents energy that is lost from each trophic level.

WARNING

Sometimes when people are talking about energy flow through an ecosystem, they say "energy is lost." If you don't think carefully about this, you might get the impression that energy is disappearing somehow. From the first law, however, we know that energy never disappears. So when someone says "energy is lost," what they really mean is "energy is lost from the system" or "energy becomes unavailable to living things." Ultimately, energy becomes unavailable for cellular work when it's transformed into heat energy because living things can't use heat to make food or fuel their cells.

Exploring energy changes

Energy can't be created or destroyed, so tracking energy through an ecosystem is all about following the flow of energy from its source to its destination. This involves two very important concepts:

- ***Energy transfer* occurs when energy is moved from a source to a receiver (also called a *sink*).** One example of an energy transfer is when electrical energy moves from the wires in your home to the battery in your cell phone.
- ***Energy transformation* occurs when energy changes from one form to another.** An example of an energy transformation occurs when electrical energy from the wires in your home transfers to a light bulb and transforms into light energy. In order to identify energy transformations in ecosystems, it's helpful to know a little bit about types of energy:
  - ***Kinetic energy* is the energy of motion.** Anything that's moving has kinetic energy. Electrical energy is a type of kinetic energy because it is based on the movement of negatively charged particles called electrons. Light energy is also kinetic energy because it consists of moving particles called photons. Moving structures like the legs of a running horse have mechanical energy.
  - ***Potential energy* is stored energy.** Specifically, potential energy is stored in objects because of their position or structure. Food contains chemical potential energy because of the structure of the chemical bonds in molecules. Tendons in the legs of animals store *elastic potential energy* when they are compressed during movement, like running. When the next step releases the compression on the tendons, they can transfer some of this energy back to the muscles to make movement more energy efficient. *Gravitational potential energy* is stored in objects held away from the surface of the Earth. Some birds that live on cliffs rely on this energy, essentially letting themselves fall off the edge to get enough speed for liftoff.

Living things transfer and transform the energy they need for their cells primarily through two major metabolic processes called photosynthesis and cellular respiration (Figure 10-5).

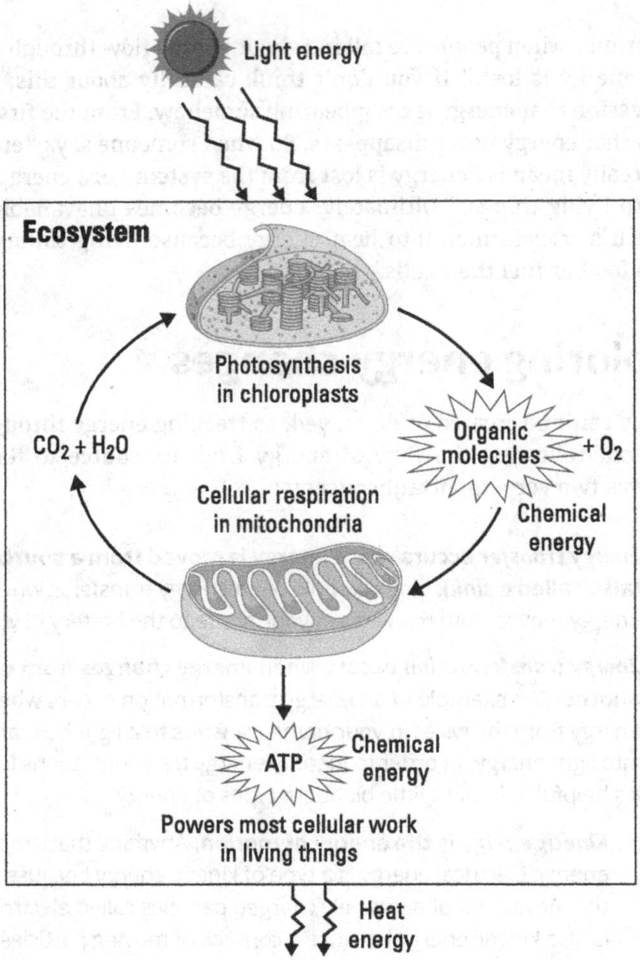

FIGURE 10-5: Energy transfers in photosynthesis and cellular respiration.

Making your own food: Photosynthesis

REMEMBER

Photosynthesis is a series of chemical reactions that occurs in the cells of photosynthetic producers, enabling them to use light energy to produce organic molecules (food) from carbon dioxide and water. Many photosynthetic producers, including plants and algae, also produce oxygen gas as waste.

Although photosynthesis occurs in many small steps, scientists usually summarize the entire process by this chemical reaction:

$$\underset{\text{carbon dioxe}}{6CO_2} + \underset{\text{water}}{6H_2O} \rightarrow \underset{\text{glucose}}{C_6H_{12}O_6} + \underset{\text{oxygen gas}}{6O_2}$$

Figure 10-5 shows an overview of how photosynthesis occurs in plants:

- » Plant cells use special organelles called chloroplasts to absorb light energy.
- » Plants absorb carbon dioxide from the air and transfer it to the chloroplast.
- » Plants use their roots to absorb water from the soil and transfer it to the chloroplast.
- » Plants use the light energy to power a series of reactions that ultimately arrange carbon, oxygen, and hydrogen atoms into the simple sugar glucose ($C_6H_{12}O_6$), which contains chemical potential energy.

Photosynthesis involves both energy transfer and transformation:

- » Energy transfer from the sun (source) to the producer (receiver or sink).
- » Energy transformation from light energy to chemical energy.

WARNING

Photosynthetic producers don't turn light energy into food. Food has substance (mass) and can be weighed because it's made of molecules. Producers need the building blocks of carbon dioxide and water to make food molecules. Light energy makes the work of photosynthesis possible.

You may be wondering where cells get other types of molecules, like fats and proteins. Once producers make glucose molecules, they can combine them with minerals like nitrogen, phosphorous, and sulfur to make other molecules needed by cells, like proteins and DNA. Consumers eat all types of food molecules when they consumer their prey. They can break down these food molecules into smaller pieces and rearrange them into the molecules they need, or build molecules from glucose with the addition of minerals.

REMEMBER

The glucose made by producers provides energy for all living things. It can also be used, in conjunction with minerals, as a building block to make other molecules that cells need. This is why producers are needed as the foundation for every ecosystem.

Releasing energy from food: Cellular respiration

Living things break down large molecules into their component parts in order to extract the energy they need. Cells chop up complex carbohydrates like starch, cellulose, and glycogen into simple glucose molecules. They release fatty acids from fats and amino acids from proteins. The small molecules from digestion can either be used as construction material for growth and repair, or they can be broken down even further to release energy for cellular work.

Most organisms on Earth use cellular respiration to release energy from food. (Many microbes and certain other cell types, like human skeletal muscle, can also use fermentation.)

REMEMBER

Cellular respiration is a complex metabolic process that cells use to break down small food molecules like glucose so they can transfer energy to power cellular work. Most cells require oxygen in order to do cellular respiration. When glucose is broken down completely, cells release the waste products as carbon dioxide and water.

Although cellular respiration occurs in many small steps, scientists usually summarize the entire process by this chemical reaction:

$$\underset{\text{glucose}}{C_6H_{12}O_6} + \underset{\text{oxygen gas}}{6O_2} \rightarrow \underset{\text{carbon dioxe}}{6CO_2} + \underset{\text{water}}{6H_2O}$$

TIP

Some ecologists refer to cellular respiration as simply *respiration*. However, because respiration also refers to breathing or gas exchange, I'll use the more specific term, *cellular respiration*, to refer to the breakdown of glucose to carbon dioxide and water.

Figure 10-5 gives an overview of the process of cellular respiration as it occurs in eukaryotic cells:

- In eukaryotic cells, cellular respiration occurs in special organelles called mitochondria.
- Cells begin breaking down food molecules, passing small fragments into the mitochondria.
- Organisms take in oxygen and transport it to the mitochondria.
- Cells harness the oxidative power of oxygen to transfer energy from food molecules to a molecule called adenosine triphosphate (ATP). Cells use ATP to power cellular work.

REMEMBER

Cellular respiration involves the energy transfer of chemical energy from food to chemical energy stored in ATP.

WARNING

Sometimes people think that producers like plants have chloroplasts instead of mitochondria. This is incorrect. Plants have chloroplasts as well as mitochondria. They use their chloroplasts to make food, but they still need mitochondria to break it down.

TIP

You can think of photosynthesis like the process of making a lunch, and cellular respiration like eating that lunch. Producers take some ingredients (CO_2 and H_2O) and make food ($C_6H_{12}O_6$). All cells (producers and consumers) unpack the lunch, using the food as a source of energy and throwing away the trash (CO_2 and H_2O).

Tracing the Flow of Energy

A continuous stream of light energy flows into Earth's biosphere. Whether or not that energy can be used to support life depends upon the work of photosynthetic producers that capture that light energy and convert it to chemical energy stored in food. (To learn more about whether light energy can support human energy needs, check out the sidebar "Shine On Me.")

REMEMBER

Primary production is the rate at which producers store energy in food. Photosynthetic producers obtain energy from sunlight, while chemosynthetic producers use energy from inorganic chemicals.

Primary production is essential for life because it provides food to ecosystems. And, because food is made from carbon dioxide and water, primary production is also an important factor in the movement of carbon through ecosystems. (See Chapter 11 for more on the global carbon cycle.) Producers remove carbon dioxide from the environment, using it to build molecules that become part of the mass of organisms.

REMEMBER

Biomass refers to the mass, measured by weight, of all living organisms in an ecosystem.

Ecologists sometimes use trophic pyramids to illustrate biomass or the numbers of individuals at each trophic level. Because of the second law of thermodynamics, all energy pyramids have the classic pyramid shape as shown in Figure 10-3, but biomass or number pyramids may have different shapes:

- ***Biomass pyramids*** **compare the dry weight of organisms in each trophic level.**
  - **Biomass pyramids for terrestrial ecosystems typically have the same shape as energy pyramids.** Producers form the largest rectangle at the base of the pyramid, and each trophic level above that gets smaller.
  - **Biomass pyramids for aquatic systems may have a producer layer that's smaller than the consumers that feed on them.** Tiny phytoplankton may experience rapid turnover in their population due to short lifespans or heavy impacts from grazing. At any given time, the amount of these producers by

weight may be small relative to the primary consumers. Still, the rapid growth of this population and high primary production ensure that enough energy is captured to support the higher trophic levels in the ecosystem.

» **The *number of individuals* in each trophic level of an ecosystem can also be shown in pyramid form.**

- **In some ecosystems, the number of individuals at each trophic level gives the same shape as an energy pyramid.** This pattern is typical in ecosystems such as grasslands that have many small individual producers, such that a large number of producers is needed to support the consumers.
- **In ecosystems that have fewer, larger producers, such as a temperate forest, the number of producers may be smaller than that of the primary consumers.** A single large individual, such as a tree, may support numerous primary consumers, such as insects, so that the number of individual primary consumers is greater than that of the individual producers.

SHINE ON ME

Scientists calculate that the sun continually delivers 173,000 terawatts (trillions of watts) of solar energy to the Earth. That's 10,000 times more energy than the global human population uses at any given time. People have been trying to figure out how to capture that solar energy since at least the early 1800s, when a French scientist named Edmond Becquerel discovered that light could generate an electric current. His discovery inspired scientists around the world to begin designing solar-powered devices, ultimately leading to the invention of the first solar cell by Charles Fritts in Edison, NJ, in 1954.

Scientists have continued to develop solar technology since the 1950s, leading to improvements in energy transfer efficiency from 1 percent in the original solar cell to an average of 22 percent in a standard silicon-based solar panel today. Recently, scientists began adding a layer of crystalline compounds called perovskites over the silicon in solar cells. Perovskites capture a broader range of light than silicon, boosting the efficiency of energy transfer from light to electricity to 27 percent (and even higher in experimental conditions). Scientists continue to experiment with other materials for absorbing light energy and new designs to best capture the light, like solar windows that may someday be installed in skyscrapers around the world.

As countries around the world scramble to find ways to slow climate change by reducing carbon dioxide emissions, many are turning to solar. China is leading the way by manufacturing more solar panels than any other country in the world. Bill McKibben, a writer and environmentalist who's been investigating climate for over 35 years, says that in May 2025, the Chinese were putting up enough solar panels per day to generate three gigawatts of power, which is the equivalent amount of power you'd expect from three coal-fired power plants. He sees other hopeful developments in solar, like the relatively inexpensive plug-in solar panels that people in Europe can simply hang on apartment balconies.

In just about eight months, citizens in Pakistan installed enough solar panels to equal half the output of the country's entire power grid. Even in the US, where the oil and gas industry gave 24 million dollars in campaign contributions in 2024, some states are making progress: California runs entirely on clean energy most days, and Utah passed a law to allow its residents to purchase small plug-in solar panels like those available in Europe.

In fact, the World Economic Forum says solar will be the dominant global power source by 2050. This progress is good news for a planet that is already 1.1°C warmer than it was before the increased use of fossil fuels that began during the Industrial Revolution. Humanity has become dependent on a large input of energy into our industrial system, and we're now in a race to supply that need while also trying to avert the worst outcomes of climate change. Rapid improvements in solar technology may help us win that race.

Inputting energy into the system

Because primary producers use some of the energy they capture to support their own functions, only a portion of the food they make is available as an energy source to other organisms in the ecosystem. Figure 10-6 shows what happens to energy as it flows into ecosystems:

- ***Gross primary productivity* is the total rate at which producers capture energy from the sun and store it in food.** It's usually measured as the rate at which carbon dioxide is captured for food production.
- **Producers use the food they make to supply energy to their own bodies and to provide building blocks for growth.** Producers use cellular respiration to release energy from food for cellular work. As their cells do cellular work, the energy from food is ultimately transferred to the environment as heat and is no longer available for other organisms in the ecosystem.
- ***Net primary productivity* is the difference between gross primary productivity and the amount of energy that's available to the next trophic level.** Net primary productivity is represented by the energy stored as biomass in the body of the producer.

Net primary productivity is expressed by the equation shown in Figure 10-6:

$$NPP = GPP - R$$

R represents cellular respiration because producers use this process to extract energy from food for cellular work.

TIP

The terminology for primary productivity is the same as what people use for paychecks. Your gross pay is how much earned based on your hours and wage before anything is deducted. Your net pay is how much you actually get to take home after all the deductions for taxes.

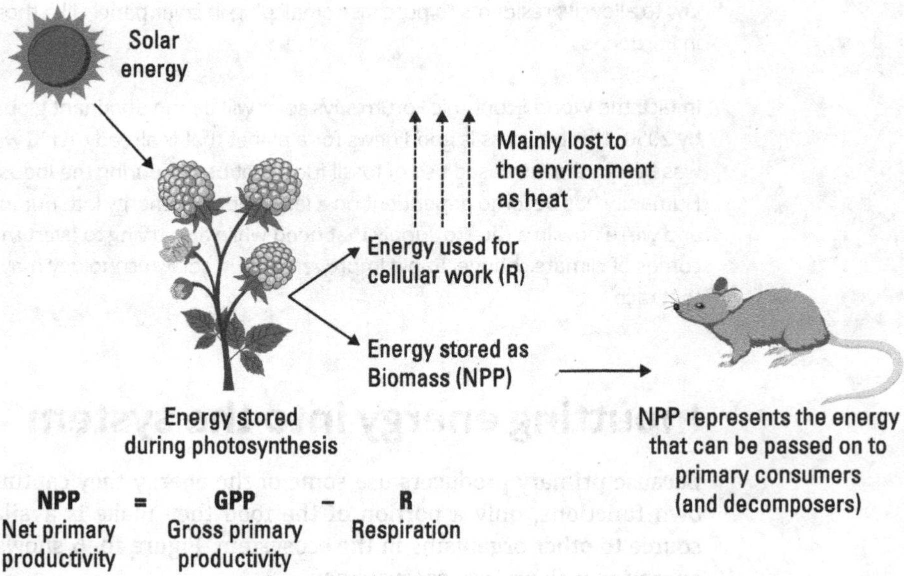

FIGURE 10-6: Net primary productivity.

Net primary production differs between ecosystems based on a number of factors:

- **Climate and length of the growing season influence NPP.** Ecosystems with warm temperatures, adequate rainfall, and long growing seasons have higher NPP. Warm temperatures and adequate rainfall favor optimal photosynthesis, and longer growing seasons result in greater plant growth. Tropical forests and savannahs have high NPP and contribute about 60 percent to total terrestrial NPP.
- **Growth in some ecosystems is limited by nutrient availability.** Although carbon dioxide is the raw material for food production, organisms must have access to other nutrients like nitrogen, phosphorus, and sulfur to build all the molecules they need for their bodies. If these nutrients are in short supply,

they may act as *limiting nutrients* that prevent growth by their absence. If producers can't grow, they can't store energy as biomass. The two nutrients that have the largest impact on most ecosystems are nitrogen and phosphorous:

» **Water availability can have direct and indirect impacts on NPP.**

- Direct impact of water on NPP occurs in ecosystems like deserts and some grasslands. Water is a reactant in photosynthesis and must be available for the process to occur. When plants are stressed by low water, they may close pores called *stomates* in their leaves to prevent water loss. This also prevents the entry of carbon dioxide into the plant, preventing it from making food.
- Indirect impacts of water on NPP may occur because a lack of water in the soil slows decomposition and limits nutrient availability. This limits growth and lowers NPP.

Runoff from agriculture or cities can introduce excess nutrients into bodies of water, removing the limitations on producer growth and increasing NPP. Unfortunately, increases in producer growth may also trigger a disastrous chain reaction that disrupts ecosystem function.

REMEMBER

Eutrophication occurs when excess nutrients like nitrogen and phosphorous accumulate in a body of water, resulting in increased algal and plant growth.

Eutrophication can cause entire lakes to turn green from the growth of algae. When the algae begin to die, the growth of decomposers increases in response to the additional food. Many decomposers use oxygen to break down food, which lowers oxygen levels in the water, sometimes resulting in massive fish kills. In addition, decomposition releases carbon dioxide, making the water more acidic and affecting shell formation in clams and other mollusks.

Keeping an eye on NPP

Ecologists track NPP over time as one measure of the health of an ecosystem. If NPP declines, it may be a sign of damage to an ecosystem from changing environmental conditions or disease. Global NPP is also very interesting to climate scientists because it is a way to measure the rate at which producers remove carbon dioxide from the environment. Scientists use a variety of methods to measure NPP, ranging from classic harvest methods to new satellite-based technologies that allow global monitoring:

» ***Harvest methods* involve collecting plant material from terrestrial systems, drying the plants to remove water, and then weighing the plants to determine their biomass.** If plant material is collected at the beginning and end of a growing season, the dry weights can be compared to determine the amount of biomass produced during the growing season. To use this method on large areas, scientists typically collect plants from smaller experimental plots within the larger area, average the change in biomass for each plot, and then scale up the number to represent the whole area.

» **The *light and dark bottle method* measures oxygen to assess the rates of photosynthesis and cellular respiration in aquatic ecosystems.** Scientists use electronic sensors to measure the amount of oxygen dissolved in the water because photosynthesis produces oxygen, while cellular respiration consumes it.

- First, scientists collect water samples from the ecosystem to be tested, and measure the amount of oxygen in the water at the beginning of the experiment.
- Next, they cover half the water bottles to block out all light. In the bottles exposed to light, organisms can do photosynthesis and cellular respiration. The dark bottles block photosynthesis, allowing only cellular respiration to occur.
- To determine NPP, scientists subtract the original amount of oxygen in the water from the amount of oxygen in the water of the light bottles at the end of the experiment. In these bottles, producers released oxygen as they captured energy from the sun, but also used oxygen as they broke down food for their own needs. Thus, whatever oxygen was added to the water represents net primary productivity.
- To determine gross primary productivity, scientists measure the oxygen in the dark bottles at the end of the experiment. The difference in oxygen in these bottles from the beginning of the experiment represents how much oxygen was used by cellular respiration. This number can be added to NPP to determine GPP.

» **Changes in carbon dioxide concentration are used to estimate NPP in terrestrial ecosystems.** Scientists use electronic sensors to measure the amount of carbon dioxide in the air because photosynthesis consumes oxygen, while cellular respiration produces it. Scientists use a combination of approaches, ranging from completely enclosing parts of an ecosystem in tents to deploying sensors at varying heights in a tree canopy. When light is available, the change in the concentration of atmospheric carbon dioxide represents NPP because it includes both the uptake of carbon dioxide for photosynthesis and the release of carbon dioxide during cellular respiration. Structures called *eddy covariance flux towers* measure how wind eddies move carbon dioxide between the producers on the Earth's surface and the atmosphere.

» **Satellite methods estimate global NPP and GPP.** Scientists use sensors on satellites to measure a variety of factors that they can use to calculate NPP and GPP. Sensors that measure chlorophyll concentration in the oceans are useful for calculating marine productivity. *Photosynthetically active radiation (PAR)* sensors measure how much light is available at the Earth's surface for photosynthesis. These sensors focus on the types of light that plants can absorb. Scientists take the data from these sensors and combine it with other environmental data, like temperature and cloud cover, to estimate how much light was absorbed for photosynthesis, allowing them to calculate GPP and NPP.

Taking energy to the next level

Primary consumers eat primary producers, breaking down some food with cellular respiration to provide energy to their bodies while also using some as raw material for growth (Figure 10-7). The energy they use for their own function is eventually transferred to the environment as heat and is unable to reach the next trophic level. Only the chemical energy that gets stored as biomass in their bodies can be passed to other organisms in the ecosystem. The amount of energy that gets stored in the body of a heterotroph is only a fraction of what was available to it from its food. (Although only a fraction of available energy gets stored in the body of a heterotroph, some dangerous molecules may accumulate as described in the sidebar "Magnifying the Problem.")

REMEMBER

The overall *transfer efficiency* of energy from one trophic level to the next is the amount of energy available in one level divided by the amount of energy available in the level immediately below it.

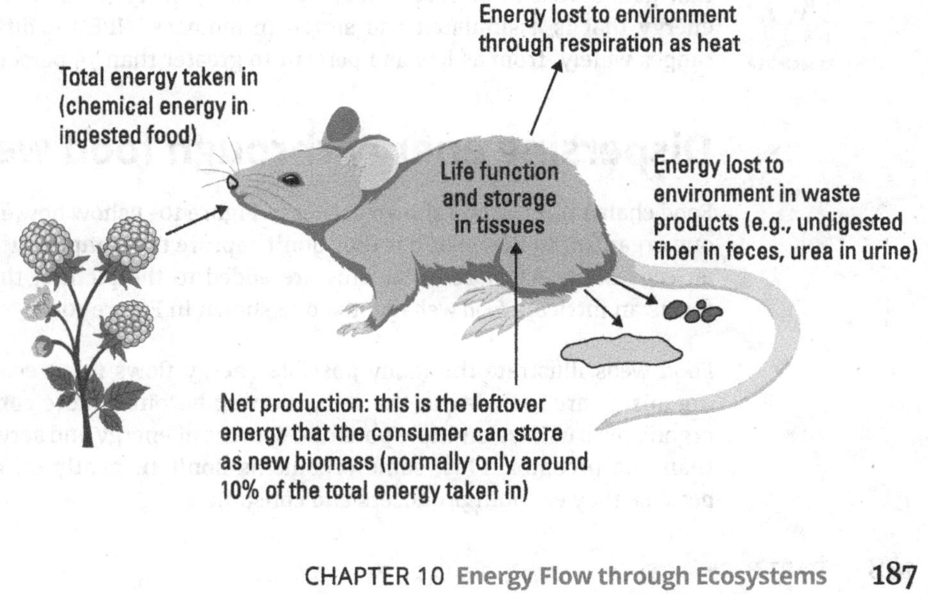

FIGURE 10-7: Net energy flow through an ecosystem.

The 10 percent rule is handy for thinking about the big picture of transfer efficiency, but the actual percentages for energy transfer differ between organisms and ecosystems. For one thing, consumers in different ecosystems may consume different amounts from the trophic level below them. For example, herbivores in terrestrial ecosystems typically consume less of the total biomass from producers than do herbivores in aquatic ecosystems. Plant defenses, like the spines on desert plants or chemicals produced in tobacco, may limit grazing by herbivores. Terrestrial plants also produce materials like wood that resist digestion by most potential grazers. Alternatively, predators on terrestrial herbivores may reduce their populations, preventing them from eating all of the available biomass.

Transfer efficiency within individual organisms can also be affected by all points along the process illustrated in Figure 10-4:

- *Consumption efficiency* reflects the fact that some consumers eat every bit of their prey, but others leave some parts behind. A snake consumes 100 percent of a rat, but orcas are known for eating just the liver from great white sharks. Consumers who don't belong to the clean plate club don't even ingest some of the available energy.
- *Assimilation efficiency* measures how much of what is eaten can be digested by the consumer. Barn owls digest small rodents whole, but then they cough up pellets that contain undigested hair and bones. Many animals and detritivores that consume a lot of plant material can't digest many complex plant fibers and end up releasing 20-50 percent of the available food energy in their waste. In contrast, carnivores can usually digest most of the material from their prey and have assimilation efficiencies of up to 80 percent.

REMEMBER

Gross production efficiency (GPE) is the percentage of the energy that was ingested that gets stored in biomass. *Net production efficiency* (NPE) is the percentage of energy that is assimilated and stored in biomass. NPE for different organisms ranges widely, from as low as 1 percent to greater than 75 percent.

Dispersing energy through food webs

Food chains like the one shown earlier in Figure 10-1 show how energy flows from one organism to the next, but they don't capture the complexity of energy flow in an ecosystem. As more organisms are added to the picture, the flow of energy forms an intricate *food web* like the one shown in Figure 10-8.

Food webs illustrate the many possible energy flows in an ecosystem. As more organisms are included in the web, the web becomes more complex. Individual organisms can have multiple potential sources of energy and serve as prey to more than one predator. Plus, some organisms don't fit neatly on one trophic level because they eat both producers and consumers.

FIGURE 10-8: A food web.

Food webs show all the possible energy connections between organisms, but they don't show the relative strength of those connections. For example, a predator may be linked to two prey organisms, but in reality may prey almost exclusively on one type of prey and only switch when that one becomes scarce. The *interaction strength* of the predator on the population of its preferred prey would probably be greater than that on the other prey species. Species that have especially strong impacts on their ecosystems are referred to as keystone species.

REMEMBER

Keystone species are species that have a greater influence on an ecosystem than their abundance would predict. In other words, their presence is a key factor in maintaining the structure and composition of that ecosystem.

Robert Paine first described the concept of keystone species in the 1960s. Paine was studying species in the rocky intertidal ecosystem of the Pacific Northwest coast. He observed that diversity in this ecosystem seemed to be higher when predators were more abundant, so he decided to conduct an experiment to test his hypothesis by removing the dominant predator from some areas and comparing them to others that he left alone.

Paine removed the predatory seastar, *Pisaster*, from his test sites, flinging them like frisbees back out into the ocean, and monitored his plots to see what would happen. The areas he left alone had diverse types of algae mixed with snails, barnacles, mussels, and, of course, *Pisaster* seastars. The plots from which he removed *Pisaster* changed dramatically. Without their predator to keep them in check, first the barnacles and then the mussels took over the rock surfaces, crowding out other species and lowering biodiversity.

Without *Pisaster*, the ecosystem changed drastically, leading Paine to describe the idea of a keystone species that was integral to its ecosystem as the top key stone in an archway is to the integrity of the arch.

MAGNIFYING THE PROBLEM

Energy isn't the only thing that passes from one trophic level to another. Chemicals known as persistent organic pollutants (POPs) can contaminate ecosystems and become concentrated in the bodies of top predators, a process known as *biomagnification.* In the Pacific Northwest, the iconic orca populations are at high risk for extinction in part due to health problems caused by POPs like polychlorinated biphenyls (PCBs) that may be present in industrial pollution.

The orcas' favorite prey are Chinook salmon, which spend much of their life eating smaller fish in the nearshore environment that may be contaminated with runoff. POPs enter the cells of producers at low concentrations, but then a primary consumer might eat large numbers of producers, effectively concentrating the POPs in their cells. A small fish could eat many of these primary consumers, concentrating the chemicals yet again. Salmon eat the small fish, and then orcas eat many salmon. The toxins become concentrated in the whale's blubber, where they may disrupt hormones and cause reproductive and neurological issues. Female whales use some of their blubber to make their milk, so the chemicals can also be transferred to new calves. The chemicals may impact growth and future fertility, as well as cause learning problems.

Some people in the Pacific Northwest are trying to decrease pollution by encouraging proper waste disposal and increased use of nontoxic products. Using public transit also cuts down on pollution from roadways, which includes harmful chemicals from tires. Unfortunately, progress in these areas may not be fast enough to save the local whale pods.

Releasing energy to space

All of the energy captured by producers and passed through food webs eventually transforms to heat, transferring first to our atmosphere and then to outer space. If you've ever become hot and sweaty during exercise, you've observed this phenomenon in action. When you exercise, your body increases the rate at which you do cellular respiration so that you can transfer more energy out of your food to fuel your muscles. Every small energy transfer produces some heat, so your body starts to warm up. The heat begins to transfer from your body to the air around you, even as your body begins producing sweat to cool you down. The heat continues to radiate away from your body and into the atmosphere.

Let's imagine you had oatmeal for breakfast and then went to the gym. Sunlight traveled from the Sun to Earth, arriving at the ecosystem where some oat plants were growing. The oats captured the sun's energy, transforming some of it to chemical energy stored as food in the oat grains. The oats also used some of the food they made for their own functions, and as they broke down food molecules with cellular respiration, some energy was transferred to heat in the atmosphere. About 10 percent of the energy that the oats originally captured was available in the grains that were harvested to make oatmeal. You ate the oatmeal, and your body assimilated some of the carbohydrates, storing them as the carbohydrate glycogen in your muscle cells.

At the gym, you begin to work out, putting extra energy demands on your muscles. Your muscle cells break down glycogen, then use cellular respiration to transfer energy for cellular work. As you do cellular respiration, some of the energy from glycogen gets transformed into heat. The work of your muscle cells transfers more energy to heat. The heat from your body transfers to the air around you and begins to make its way to space. In summary, energy from the sun flowed into the oats, then into you, and then made its way back to outer space as heat.

REMEMBER

Ecologists say that energy flows through ecosystems.

Your body doesn't just process the energy from your food; it also recycles the atoms that make up the mass of the food you eat. You can read about matter recycling by heading to Chapter 11. The presence of greenhouse gases in our atmosphere slows the passage of heat to outer space. Without any greenhouse gases, Earth would be too cold to support life as we know it. The increase in greenhouse gases since the Industrial Revolution is slowing the passage of heat to a point where the global temperature is rising and putting stress on ecosystems. (See Chapter 12 for more on the Greenhouse Effect and global warming.)

IN THIS CHAPTER

» Discovering what you're made of

» Finding out some chemistry fundamentals

» Exploring recycling on a global scale

Chapter 11 Recycling Matter through Ecosystems

All living things are made of the same basic elements, mostly carbon, hydrogen, oxygen, nitrogen, phosphorus, and sulfur. Organisms must have a source of these elements, either directly from the environment or by eating others. As organisms grow and reproduce, they release some elements back to the environment as waste. Eventually, organisms die, and decomposers recycle the remaining material. The movement of matter through ecosystems is an amazing game of transformation as elements shift forms and combine in new ways. For example, one moment, a carbon atom could be part of a protein in the toe of a frog, the next could be released to the air as carbon dioxide by a hawk that ate the frog. Then a plant could absorb the carbon dioxide and use it to build a sugar molecule. Nothing is wasted, and nothing is lost. When one organism throws something away, another is there to pick it up. In this chapter, you'll see how organisms participate in the global recycling of the elements needed for life.

Identifying the Building Blocks of Life

Have you ever heard the quote "We are made of star-stuff," and wondered what it meant? Carl Sagan said this back in the 80s on his TV show *Cosmos*. Part of what he was saying is that the basic elements that make up our bodies — the hydrogen, oxygen, carbon, calcium, iron, and other "stuff" — were formed as part of the life cycle of stars. Then, when massive stars reach the end of their lives, they explode into supernovae, scattering elements into space. These elements may become part of the dust and gas that orbit new suns and eventually collect into masses that collide with each other, forming planets and moons. So, the elements that make up planet Earth, the ones that form the bodies of all living things, including you, were originally formed in a star.

What makes matter?

The world around you, nonliving and living, is made of atoms and molecules. Atoms are the smallest units of matter that have the unique properties of elements.

REMEMBER

Elements are pure substances that can't be broken down into other things by chemical means.

TIP

You can think of elements as the flavors of matter, like iron, neon, sodium, and helium. If you've ever seen a periodic table of the elements, you've seen a chart that shows the many different flavors of matter.

So far, scientists have discovered 118 different elements, but only a small fraction of these make up the cells of living things. Based on the structure of their atoms, each element has unique characteristics that affect its physical properties and how it can combine with other elements. Scientists abbreviate the different elements with one or two letters. For example, you're probably familiar with H for hydrogen and O for oxygen.

REMEMBER

Nutrients are the elements that organisms require for their metabolism and growth.

Organisms need some nutrients in large amounts because they form the bulk of structures in cells. These *macronutrients* are carbon (C), hydrogen (H), nitrogen (N), oxygen (O), phosphorus (P), and sulfur (S). Cells also need relatively large amounts of magnesium (Mg), calcium (Ca), sodium (Na), and potassium (K), as well as smaller amounts of many elements like iron (Fe) and zinc (Zn). Scientists call these elements *micronutrients* or *trace elements*.

Based on their properties, some atoms attach to each other by attractions called *chemical bonds*, forming larger structures called molecules. The large molecules, or *macromolecules*, that form the structures of living things are proteins, carbohydrates, lipids like fats and oils, and nucleic acids like DNA.

WARNING

In nutrition and fitness, people often refer to *macros* or *macronutrients* as the type of foods our bodies need in the greatest amounts. In this context, macros refer to proteins, carbohydrates, and fats.

To build their cells, living things need sources of their essential nutrients. Heterotrophs get what they need by eating others, while autotrophs obtain matter from the environment. (To learn more about the different types of heterotrophs and autotrophs, head back to Chapter 10.)

Matter vs. energy

Matter is the stuff that life and the Earth are built from, the things you can touch or feel. Even the air, or atmosphere, is made of matter: it may seem like nothing, but if you've ever stuck your arm out of the window of a moving car, you've felt the molecules of gas in the air pushing on your arm. Scientists define matter as anything that has *mass*, in other words, anything that can be weighed. All living things need a source of matter to grow and repair their cells.

In addition to matter, living things need energy to do the work of cells. They need energy to move, to build molecules, and to maintain their organization. Energy may be stored in matter because of the arrangement of atoms, but energy itself has no mass and can't be weighed. For example, you can't hold light or electricity in your hand.

REMEMBER

Energy is anything that can do work or cause a change.

TIP

Einstein's theory of special relativity, $e = mc^2$, says that matter moving at the speed of light squared is equal to its kinetic energy, but this doesn't happen in living things. So, in ecosystems on Earth, matter and energy behave as two separate things.

Energy can move from one place to another and can be changed from one form to another (for more about energy transfer and transformation, flip back to Chapter 10). Organisms use energy constantly to maintain themselves, and without energy, they would die.

Tracing Elements through Biogeochemical Cycles

Star staff provided the matter to form the Earth, and that's pretty much all the matter that life will ever have available to it. With the exception of the occasional rocket that leaves the Earth or meteorite that crashes into it, the amount of matter available on Earth is fixed. Even the molecules in our upper atmosphere don't leak out into space. What we've got is all we'll ever have.

This is when we have to pause and thank the decomposers, detritivores, and scavengers (head back to Chapter 10 for more details on these groups). If it weren't for the organisms that break down molecules in the dead, life might have run out of matter long ago. Fortunately, when organisms die, other organisms digest the matter that makes up their cells. The dead become the food for the living, recycling matter back to ecosystems. Decomposers incorporate some molecules into their own cells as building blocks and break others down for energy, releasing waste products like carbon dioxide back to the atmosphere. Autotrophs like plants use the carbon dioxide to produce new food molecules, and the cycle begins again.

REMEMBER

Ecologists say that matter cycles through ecosystems.

Elements move from the environment to organisms and then back again. In the environment, elements are part of *inorganic molecules*, those that aren't found in living things. Organisms chemically shuffle elements around to form the organic molecules that build cells.

REMEMBER

Organic molecules contain carbon and are found in living things.

As matter cycles through ecosystems, the elements are continuously recycled from inorganic to organic molecules and then back again.

Redox reactions: Playing catch with electrons

Organisms change molecules from one form to another through a series of chemical reactions called *metabolic pathways*. Metabolic pathways are intricately linked with the flow of energy in cells. In particular, energy transfer in cells happens through oxidation and reduction reactions:

- **Molecules lose electrons during *oxidation reactions*.** In Figure 11-1, molecules A and B react with each other. The figure shows the molecules before the reaction on the left, then after the reaction on the right. During the reaction, molecule A is oxidized as it gives electrons to molecule B.
- **Molecules gain electrons during *reduction reactions*.** In Figure 11-1, molecule B is reduced as it accepts electrons from molecule A.

Oxidation and reduction reactions always occur together because the electrons move from one molecule to another: one is oxidized and the other reduced. Scientists refer to these paired reactions as *redox reactions*.

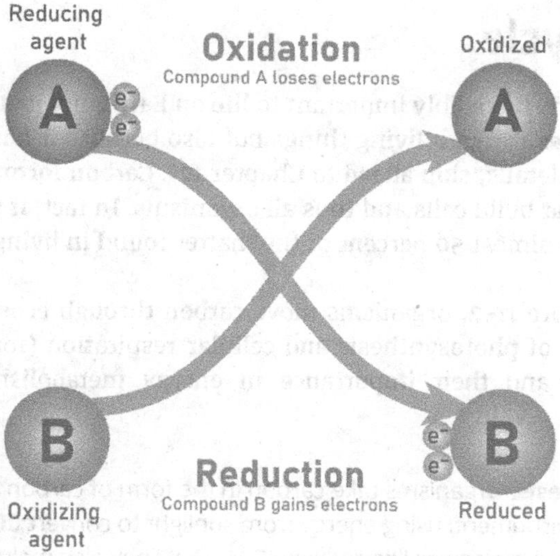

FIGURE 11-1: Oxidation-reduction (REDOX) reactions.

SAMYA/Adobe Stock Photos

TIP

You can think of redox reactions like a game of catch. The molecule that gives up electrons is like the person throwing the ball. The molecule that receives electrons is the person catching the ball.

Redox reactions are essential for the transfer of energy in cells. Oxidation reactions release energy to cells, while reduction reactions require energy. Cells oxidize energy-rich molecules to obtain the energy they need for cellular work, but they store energy by reducing molecules. Because of the second law of thermodynamics (explored in Chapter 10), some of the available energy is also transformed into heat with every transfer. The heat transfers into the environment and is no longer useful as an energy source to cells.

TIP

If the terms oxidation and reduction seem backward to you, you're not alone. One way to keep the reactions sorted out is to remember the phrase "LEO the lion goes GER." This phrase can remind you that Loss of Electrons is Oxidation (LEO), while Gain of Electrons is Reduction (GER).

As living things oxidize and reduce molecules as part of their metabolism, they constantly release waste materials. But what is trash to one organism may be treasure to another. Organisms pick up these molecules, use them for their metabolism, and release them in a new form. Scientists map the transformation of elements through ecosystems, focusing on each of the macroelements one at a time.

REMEMBER

Biogeochemical cycles track the movement of elements (*chemo* = chemical) through living things (*bio* = living) and back to the physical environment (*geo* = earth).

Carbon cycle

The carbon cycle is incredibly important to life on Earth, not just because it shows how carbon moves through living things but also because it has a big impact on our climate (for details, skip ahead to Chapter 12). Carbon forms the backbone of the molecules that build cells and thus all organisms. In fact, if you ignore water, carbon makes up almost 50 percent of the matter found in living things.

As shown in Figure 11-2, organisms move carbon through ecosystems primarily by the processes of photosynthesis and cellular respiration (for more details on these processes and their importance in energy metabolism, head back to Chapter 10):

- **Photosynthesis:** Organisms take carbon in the form of carbon dioxide (CO_2) from the environment, using energy from sunlight to convert CO_2 and water (H_2O) into food molecules like glucose ($C_6H_{12}O_6$). They also make oxygen gas (O_2) as waste. Figure 11-2 shows the role of the tree and the grass in moving CO_2 out of the environment and into living things, where it becomes organic molecules such as proteins, carbohydrates, and fats.
- **Cellular respiration:** Organisms use oxygen to extract energy from carbon-containing food molecules, releasing CO_2 back to the environment. They also produce H_2O as waste. Figure 11-2 shows that respiration by plants, animals, and decomposers moves carbon out of living things and back into the environment as CO_2.

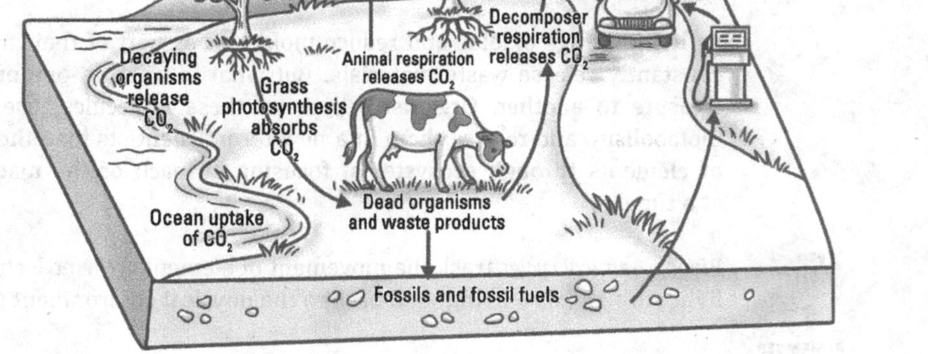

FIGURE 11-2: The global carbon cycle.

REMEMBER

Photosynthesis reduces CO_2 to $C_6H_{12}O_6$; cellular respiration oxidizes $C_6H_{12}O_6$ back to CO_2.

The carbon cycle is a great example of how one organism's trash is another organism's treasure. You breathe out carbon dioxide (CO_2), which is a very oxidized form of the element carbon that has no further value to your metabolism. But to a plant, CO_2 is an essential building block for the construction of food that can be used to build the molecules of all the organisms in the ecosystem. As organisms live, grow, and die, carbon keeps cycling from the environment to the molecules of cells and then back again.

REMEMBER

When organisms die, decomposers break down their carbon-containing molecules for energy, releasing the carbon back to the environment as CO_2.

WARNING

The biggest mistake people make about the carbon cycle is to think that carbon moves from dead things to the soil and then back to plants. It's easy to see why people might assume this: we see things die and decompose until they seem to disappear into the soil. We see plants grow, and we know they have roots anchored in the soil. But the reality, the part we don't see, is that the dead disappear as decomposers digest their molecules, using them to grow their own cells and provide energy. When decomposers use cellular respiration to get energy from the dead, they release carbon back to the environment as CO_2 (which is a gas and therefore invisible). So, when dead things seem to break down, it's partially because they're disappearing into thin air! Plants recapture carbon as CO_2 from the air, then build their bodies by rearranging the carbon from these invisible gas molecules into the backbones of their proteins, carbohydrates, and fats. Plants indeed take water and nutrients in through their roots, but these nutrients are small amounts of minerals that provide nitrogen, phosphorus, sulfur, calcium, and other elements. Yes, these atoms get built into the plant body, but the main plant body, the vast bulk of it, is built from molecules that have carbon backbones. So, the dry mass of a plant mostly comes from the atoms in CO_2 molecules. The real story of the carbon cycle is that carbon moves from dead things to the air and then back to plants.

Another major component of the carbon cycle is the release of CO_2 from the burning of fossil fuels. Fossil fuels like coal and oil formed from the remains of living things that were deposited in low-oxygen environments millions of years ago. Without oxygen, decomposers in these environments didn't fully digest the carbon-containing molecules in the dead. Sediment settled over the remains, and then heat and pressure over millions of years gradually converted them to carbon-rich fossil fuels. When people burn these fuels, the carbon returns to the environment as CO_2. (Chapter 12 explores how the increased burning of fossil fuels since the Industrial Revolution is affecting our climate.)

Several other processes also contribute to the movement of carbon through ecosystems:

- Volcanoes vent carbon dioxide from underground magma and can release large amounts of this gas during eruptions.
- Forest fires release carbon dioxide as they rapidly oxidize the molecules of the living things they consume.
- Just like photosynthetic autotrophs, chemosynthetic autotrophs remove carbon dioxide from the environment as they build food molecules. Because they use chemicals as the source of energy for this process, chemoautotrophs can grow in ecosystems without light, such as the bottom of the ocean or in lightless caves.
- Some dead organic matter gets buried in sediments. Over long periods of time, pressure and heat may convert this material to sedimentary or metamorphic rocks or even to fossil fuels.

Each of these processes moves carbon from one form to another through Earth's ecosystems. Some processes lock carbon into organic molecules in living things or inorganic molecules in rocks, while other processes move carbon back into the environment as carbon dioxide. The relative speed of these processes determines the overall balance of carbon in the environment. And, because carbon dioxide is a *greenhouse gas* that can trap heat in the atmosphere, the carbon cycle has huge implications for Earth's climate (discussed further in Chapter 12).

Nitrogen cycle

All life needs nitrogen to build molecules like proteins and DNA. The largest reservoir of nitrogen on Earth is our atmosphere, which is almost 80 percent nitrogen gas (N_2). This is more than enough nitrogen for all life on Earth, but there is one catch: most organisms have no way to access this form of nitrogen. For example, you draw nitrogen gas into your lungs with every breath you take, but your body can't use it, so you just breathe it out again. You get your nitrogen from molecules like protein in the food you eat, but those organisms had to get their nitrogen from somewhere, too. Ultimately, nitrogen from the atmosphere must be converted into forms of nitrogen that producers can use and incorporate into their cells. These producers then form the basis for food webs that spread nitrogen to all living things in the ecosystem (for more on producers and food webs, check out Chapter 10).

REMEMBER

Nitrogen fixation is the conversion of atmospheric nitrogen (N_2) into nitrogen-containing compounds that living things can use.

Figure 11-3 shows the fixation of nitrogen by both nonliving and living things:

- **Abiotic nitrogen fixation** occurs when nonliving forces like lightning provide enough energy for N_2 to react with oxygen, forming compounds like nitric oxide (NO), nitrite (NO_2^{2-}), and nitrate (NO_3^-). Rain carries these compounds into the soil, where plants and microbes can absorb them.
- **Biological nitrogen fixation** occurs by the action of a few special groups of microbes. These microbes use cellular energy and enzymes called nitrogenases to convert N_2 into ammonia (NH_3):
  - Cyanobacteria (a.k.a. blue-green algae) are common in aquatic environments and soil.
  - Nitrogen-fixing bacteria may live freely in the soil or form symbiotic associations with plant roots (for more on symbiosis, head back to Chapter 7). Bacteria called rhizobia are particularly important in agriculture because they associate with plants in the legume or bean family. These bacteria cause the plant roots to grow little projections, called root nodules. The bacteria move into the nodules where they grow and fix nitrogen. Some of the ammonia they produce is funneled to the plants, improving their growth. Farmers will sometimes plant cover crops of legumes like alfalfa, then plow the alfalfa into the field to increase the nitrogen content in the soil. For an example of the importance of nitrogen fixation after disturbance, head back to Chapter 9.
  - Some prokaryotes in the domain Archaea also live in the soil and fix nitrogen.

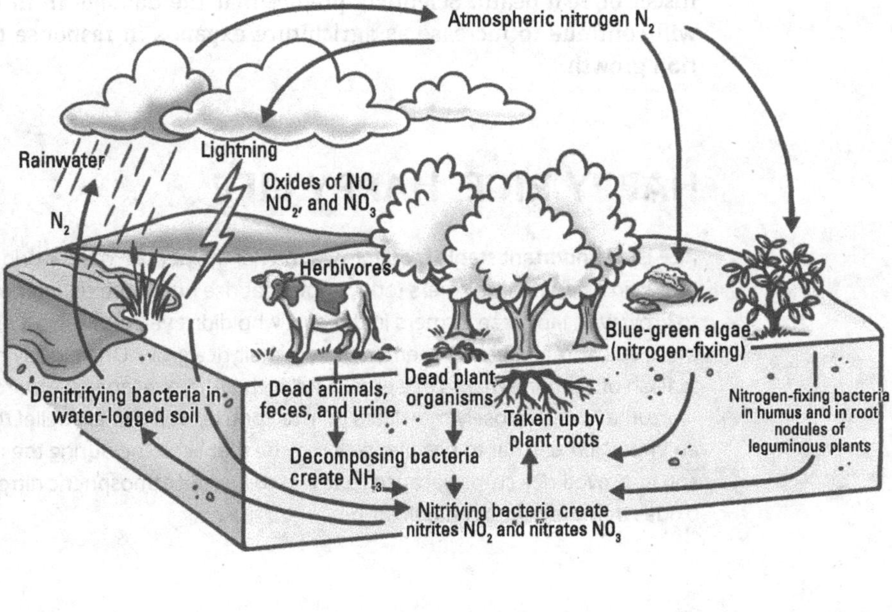

FIGURE 11-3: The global nitrogen cycle.

Nitrogen fixation delivers nitrogen to water and soil in forms that can be used by cells. This valuable nitrogen is taken up quickly by cells and doesn't usually remain in the environment for very long. Because of this, humans often supplement crops with chemical fertilizers. Most chemical fertilizers are manufactured by an industrial process that uses energy to combine nitrogen gas (N_2) with hydrogen gas (H_2), producing ammonia (NH_3). The ammonia may then be reacted with nitric acid to produce ammonium nitrate, which plants easily absorb. The sidebar "Happy Rice, Happy Life" describes how fertilizer may be produced naturally during storms.

Scientists estimate that the production of fertilizer for agriculture has doubled the global rate of nitrogen fixation. Excess usable nitrogen enters ecosystems and groundwater, resulting in harmful outcomes like eutrophication (Chapter 10), loss of biodiversity (Chapter 13), toxic algal blooms, and acidification of freshwater ecosystems. Nitrates in human drinking water can cause blue baby syndrome, which affects the ability of red blood cells to carry oxygen and can be fatal, and may also increase the risk for certain cancers.

REMEMBER

Some forms of nitrogen pollution contribute to the production of *acid rain*. Acid rain, or *acid deposition*, refers to precipitation like rain, snow, fog, sleet, and dew that has a pH value lower than that of normal precipitation.

Nitrogen oxides in the atmosphere combine with water vapor to become nitric acid (HNO_3), which returns to the Earth's surface as acid rain. The sources of nitrogen oxides include coal-fired power plants, vehicle engines, and ammonia vapors released by agricultural operations. These atmospheric pollutants can travel long distances from their source before they fall as rain. Acid rain affects nutrient availability in the soil, damaging plant growth and sometimes leading to insect or fish death. Scientists predict that the damage from nitrogen pollution will continue to increase as agriculture expands in response to human population growth.

HAPPY RICE, HAPPY LIFE

Rice is an important staple food for over half of the people in the world. Most rice is grown in Asia, where farmers today know that rice needs the right amount of nitrogen to grow well. Japanese farmers in the past who didn't yet know about nitrogen knew there was something that seemed to help their rice grow. One old saying translates to "a flash of lightning makes rice grow a full inch." And one Japanese word for lightning is "inazuma," which loosely translates to "rice spouse," reflecting a belief that lightning was an important partner to rice plants. It may be that lightning during the rice growing season improved rice crops because lightning converted atmospheric nitrogen into nitric oxides that the plants could absorb.

Once usable nitrogen gets delivered to ecosystems, the nitrogen cycle can really begin. Organisms take up the various forms of nitrogen, use them in their metabolism, and then release new forms of nitrogen that can be grabbed and used by somebody else. Nitrogen keeps changing forms, and the whole cycle can seem very complicated. If you sort it all out, however, it really comes down to four main processes:

- **Assimilation is the formation of organic nitrogen compounds like proteins and DNA from inorganic sources of nitrogen in the environment.** Plants use their roots to absorb ammonia and nitrate along with water from the soil, then incorporate the nitrogen atoms into their molecules. Likewise, aquatic producers assimilate nitrogen from their environments. The nitrogen assimilated into the bodies of producers then spreads through food webs as organisms eat other organisms.
- **Ammonification is the release of ammonia to the soil as a waste product of decomposition.** As decomposers digest the proteins of the dead, they remove the nitrogen-containing groups from the amino acids. If the decomposers have nitrogen for their own molecules, they will release the excess as ammonia into the soil. Ammonification also occurs from the breakdown of animal wastes such as urine.
- **Nitrification is the production of nitrates from the oxidation of ammonia.** *Nitrifying microbes* can use reduced forms of nitrogen, such as ammonia (NH_3), as a source of chemical energy. These microbes oxidize ammonia to produce nitrite (NO_2^{2-}). Other microbes can further oxidize the nitrite, producing nitrate (NO_3^-).
- **Denitrification removes nitrates from the soil by reducing them to nitrogen gas.** *Denitrifying microbes* can use oxidized forms of nitrogen instead of oxygen during cellular respiration. These microbes may reduce nitrate (NO_3^-) to nitrite (NO_2^{2-}), nitrous oxide (N_2O), or nitrogen gas (N_2). Denitrification results in the return of some N_2 to the atmosphere, closing the loop in the nitrogen cycle.

TIP

Plants can easily access nitrogen when it's present as nitrates in the soil. So, if you think about the nitrogen cycle from the plant's perspective, nitrification adds the plant's favorite form of nitrogen to the soil, while denitrification removes it.

TIP

You can look at the number and types of atoms in a nitrogenous compound for clues about the oxidation state of the nitrogen atom. The more oxygen atoms you see in the compound, the more oxidized the nitrogen. For example, NO_3^- is more oxidized than NO_2^{2-}, which is more oxidized than N_2. When you see bonds with hydrogen, the nitrogen is more reduced. For example, NH_3 is more reduced than N_2. Oxidized forms of nitrogen are useful as electron acceptors to microbes that do anaerobic cellular respiration. Reduced forms of nitrogen are sources of chemical energy to chemosynthetic autotrophs.

Phosphorus cycle

Organisms need phosphorus for essential phosphate-containing cellular components, including DNA and phospholipids. The largest reservoirs of phosphorus on Earth are in rocks and ocean sediments. Figure 11-4 shows how weathering of rocks and sediment releases usable phosphorus into the environment as phosphate (PO_4^{3-}), making it available to living things:

- » Phosphate enters soil and water, where plants and aquatic producers such as phytoplankton absorb some of it.
- » Producers *assimilate* the phosphate, incorporating it into their organic molecules such as DNA, RNA, and the energy carrier ATP.
- » Phosphate moves through the food web to all types of organisms through processes like herbivory, predation, and decomposition.
- » Decomposers release excess phosphate back into the upper soil environment, where it is accessible to plant roots.
- » As unassimilated phosphate sinks lower into soil and sediment, it reacts with other minerals and becomes unavailable to living things.

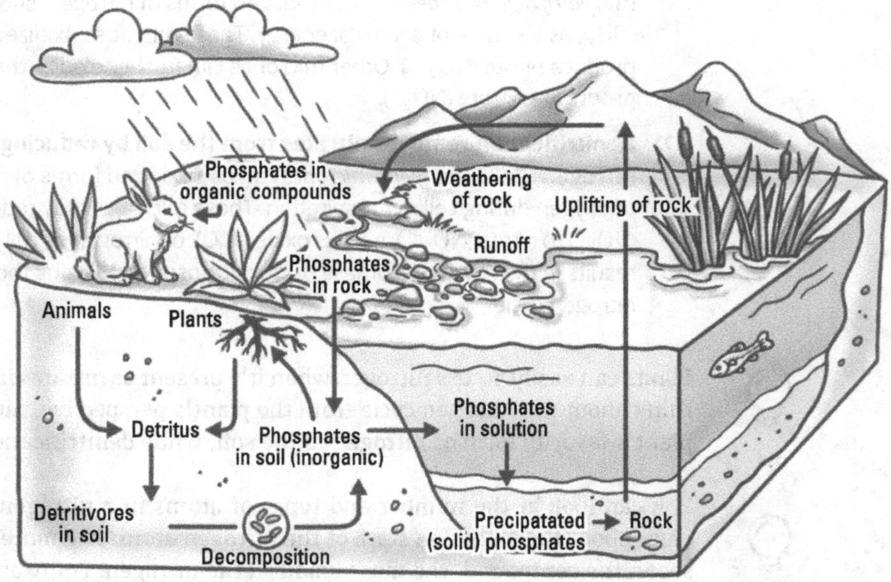

FIGURE 11-4: The global phosphorus cycle.

The weathering of rocks releases phosphorus very slowly into the environment. Humans have greatly increased this flow of phosphorus by mining for phosphorus to use in the production of fertilizers and products such as detergents. When agricultural runoff and phosphate-containing wastewater enter aquatic ecosystems, they can trigger massive algal blooms and eutrophication of lakes (Chapter 10). The damage to lakes such as the Great Lakes in the United States led to the removal of phosphates from most detergents in the 1990s. Since then, many states and countries have passed laws banning phosphates from laundry and dishwashing detergents.

Sulfur cycle

Organisms need sulfur because it's part of many proteins and cofactors. The largest reservoir for sulfur is the Earth's crust, where it is part of sedimentary rocks. Sulfur is also abundant as sulfate (SO_4^{2-}) in ocean water and marine sediments. Sulfur moves through ecosystems by both abiotic and biotic processes, changing from highly oxidized forms like sulfate (SO_4^{2-}) to very reduced forms like hydrogen sulfide (H_2S) and many forms in between. The fact that sulfur has so many different oxidation states makes it particularly useful for microbial metabolism.

Figure 11-5 shows how sulfur moves through ecosystems:

- Sulfur compounds enter ecosystems from reservoirs in the Earth's crust:
  - Volcanoes release sulfur dioxide (SO_2) into the atmosphere.
  - Spray from the ocean and dust blown from the land release sulfur to the atmosphere.
  - Inorganic, sulfur-containing minerals and rocks release sulfate (SO_4^{2-}) into rivers and oceans.
  - Hydrothermal vents on the ocean floor release hydrogen sulfide (H_2S).
- Abiotic processes move sulfur through ecosystems:
  - Sulfur dioxide in the atmosphere reacts with oxygen and water vapor to become sulfuric acid (H_2SO_4), which falls to the earth as part of acid precipitation.
  - Rivers carry sulfates into the oceans.
  - Sulfur-containing compounds may become buried under layers of sediment. Over long periods of time, heat and pressure can convert these compounds into rocks.

» Organisms move sulfur through ecosystems:

- Plants absorb sulfate from the soil, reducing it and assimilating it into their proteins. These proteins move through the food web to all types of organisms.
- Decomposers release excess sulfur as hydrogen sulfide (H_2S) to the environment.
- *Chemoautotrophs* can use reduced sulfur compounds like H_2S as a source of chemical energy. These *sulfur-oxidizing* microbes take electrons from reduced forms of sulfur, producing more oxidized forms like sulfate (SO_4^{2-}) as waste.
- Some *photoautotrophs*, like the green and purple sulfur bacteria, can use hydrogen sulfide (H_2S) instead of water (H_2O) as a source of electrons during photosynthesis. These bacteria oxidize H_2S, producing S_0 as waste. Because this type of photosynthesis doesn't produce oxygen gas (O_2), it's called *anoxygenic photosynthesis*. (For details on oxygenic photosynthesis, flip back to Chapter 10.)
- *Sulfate-reducing* microbes can use oxidized sulfur compounds like sulfate (SO_4^{2-}) instead of oxygen (O_2) as an electron acceptor during cellular respiration. As these microbes perform *anaerobic cellular respiration*, they produce reduced sulfur compounds like hydrogen sulfide (H_2S) as waste.

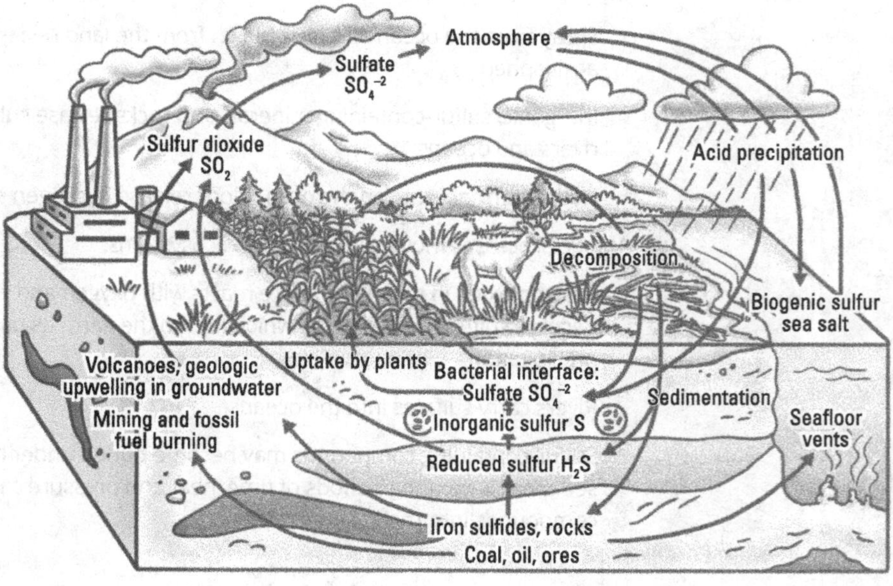

FIGURE 11-5: The global sulfur cycle.

TIP

Just as you can for nitrogenous compounds, you can look at the number and types of atoms in a sulfur compound for clues about the oxidation state of the sulfur atom. The more oxygen atoms you see in the compound, the more oxidized the sulfur. For example, SO_4^{2-} is more oxidized than sulfur dioxide (SO_2), which is more oxidized than elemental sulfur (S_0). When you see bonds with hydrogen, as in hydrogen sulfide (H_2S), the sulfur is more reduced. Oxidized forms of sulfur are useful as electron acceptors to microbes that do anaerobic cellular respiration. Reduced forms of sulfur are sources of chemical energy to chemosynthetic autotrophs and sources of electrons to microbes that do photosynthesis without producing oxygen.

Human impact on the sulfur cycle results primarily from the mining and burning of coal, which naturally contains sulfur. Sulfur-oxidizing bacteria access sulfides in coal mine drainage, producing sulfates that combine with water to produce sulfuric acid. Acidic mine drainage leaches into natural waters, killing plants and animals and potentially making drinking water unsafe. People are remediating the land around abandoned mines with a variety of strategies, including the addition of chemicals to balance the pH and harnessing the power of microbes to clean up the toxic metals in the water.

The burning of coal releases sulfur dioxide into the atmosphere, where it reacts with oxygen and water vapor to produce sulfuric acid. When people started burning more coal during the Industrial Revolution, they began adding more sulfur dioxide to the atmosphere, resulting in increased acid precipitation. Sulfur dioxide emissions continued to increase until people started calling for change to stop the devastation to forests and damage to national monuments. In the United States, the Clean Air Act of 1970 required coal-fired power plants to install sulfide scrubbers, leading to a reduction in sulfide emissions and a decrease in sulfur-based acid precipitation.

The sulfur cycle has an important connection to climate control. Sulfates produced in the atmosphere during the sulfur cycle can become part of *cloud condensation nuclei*, small atmospheric particles that encourage cloud formation. Clouds reflect sunlight out to space, keeping the planet cooler. In addition to sulfur dioxide emissions, sulfate can form from an organic molecule, dimethyl sulfate (DMS), produced by marine phytoplankton. Some scientists have proposed that the link between phytoplankton and cloud formation might form a natural feedback mechanism to help regulate the planet's temperature (see the sidebar The Gaia Hypothesis for more details).

THE GAIA HYPOTHESIS

In the 1960s, when humans were first beginning to explore space, NASA asked the global scientific community how they would look for life on other planets. In other words, what kinds of evidence should NASA collect? Photographic evidence didn't seem to be the answer — photos from some areas of the Earth looked just as lifeless as photos taken by probes on the surface of Mars and Venus. A British chemist named James Lovelock thought about this problem and came up with an answer that then led him to a revolutionary idea that he named the Gaia Hypothesis.

Dr. Lovelock's answer to NASA's question was a chemical one. He said you could tell if a planet is alive by looking at the chemicals in the lower atmosphere. The atmospheres of our nearest neighbors, Mars and Venus, are mostly carbon dioxide with small amounts of nitrogen gas. But Earth? Our atmosphere is mostly nitrogen, with lots of oxygen gas, and small amounts of a variety of very reactive and unstable substances that can't be explained by chemistry alone. If there were no life on Earth, the chemicals in our lower atmosphere would react and reach equilibrium, producing an atmosphere very similar to those of Mars and Venus. The reason our atmosphere is so different is that living things participate in biogeochemical processes that constantly release unstable products into the atmosphere.

Then Dr. Lovelock took his idea one step further. He wondered if it was possible that the living things on Earth interacted with the environment in a way that maintained conditions so that life could exist. In other words, he wondered whether the biosphere itself functioned much like a human body with feedback systems that worked to keep the biosphere alive. Just like you might start sweating on a hot day to cool your body down, could warm temperatures on Earth promote the growth of phytoplankton so that they produce more dimethyl sulfate? Once in the atmosphere, dimethyl sulfate can become part of cloud condensation nuclei, increasing cloud formation. Clouds reflect heat to space, helping keep the planet cool.

Dr. Lovelock called his idea the Gaia hypothesis. At its core is a vision of the biosphere as one giant, self-regulating system. Dr. Lovelock worked with microbiologist Lynn Margulis to develop his hypothesis, looking for evidence of how life transformed the environment of the early Earth into what we experience today. Although the Gaia hypothesis is controversial, scientists have discovered many mechanisms by which living things impact environmental factors such as global temperature, ocean salinity, and atmospheric composition.

4

Looking at the Big Picture of Global Ecology

IN THIS PART . . .

Understand global climate change

Find strength in diversity

Explore the practical applications of ecology

IN THIS CHAPTER

» Distinguishing climate variation from climate change

» Examining the evidence for human impacts on climate change

» Predicting the consequences of climate change

Chapter 12 Global Climate Change

One of the greatest challenges people face today is the impact that humans are having on our global climate. Average global temperatures have climbed steadily since the 1960s, and recent years have been some of the hottest on record. The changing climate has triggered unusual weather patterns, including increasing numbers of severe storms. Climate determines where organisms can live and how they function (see Chapter 2 for details), so climate change has the potential to completely reshape Earth's ecosystems. In this chapter, you will consider the evidence, causes, and consequences of our changing climate.

Causing Climate Change

You probably notice the weather every day: whether it's hot or cold, sunny or rainy. You're also probably aware of the general climate in your area: how hot it usually gets during the summer or which season is the wettest. Weather and climate both change naturally, although people are usually more aware of the short-term changes in weather than they are of longer-term changes in climate (for details on the differences between weather and climate, head back to Chapter 2). Some people do notice variations in climate, like a series of warmer winters or wetter summers.

REMEMBER

Climate variation occurs when the climate patterns change in time or space. In terms of time, climate may vary over decades, centuries, millennia, or even longer periods like ice ages. In terms of space, climate can vary locally, regionally, or even globally.

Climate variation is different from climate change. Climate variation refers to the general fluctuations in climate around the average. Climate variation leads to changes from year to year, but the overall climate remains consistent.

REMEMBER

Climate change refers to long-term continuous changes in climate averages.

Climate scientists typically use 30-year averages of climate factors like rainfall and temperature as *climatological normals* that represent a climate. So, scientists typically recognize climate change when averages shift in one direction, either increasing or decreasing, for 30 years or more.

REMEMBER

Scientists agree that Earth is currently undergoing climate change. According to the American Academy for the Advancement of Science (AAAS), 97 percent of scientists conclude that these changes result from human actions.

REMEMBER

Global warming refers to the steady increase in the global annual temperature. The planet's average temperature has risen about 2 degrees Fahrenheit (1 degree Celsius) since the late 19th century. The World Meteorological Society (WMO) reports that 2025 was one of the three warmest years on record and that the past 11 years have been the 11 warmest on record.

Climatologists who study global warming often use the average temperature from 1850-1900 as the climatological normal to represent Earth's temperature before the Industrial Revolution. They report on temperature changes as *anomalies* from that average. For example, the average global temperature reported by the National Oceanic and Atmospheric Administration (NOAA) for 2024 was approximately 15.2 degrees Celsius. The pre-industrial climatological normal is approximately 13.7 degrees Celsius. Climatologists would report this as a +1.54 degrees C anomaly, meaning that the average global temperature in 2024 was 1.54 degrees warmer than in the pre-industrial period.

Here comes the sun

The Sun is the source of most of the energy for life on Earth. Energy arrives as electromagnetic radiation from the sun, flows through our atmosphere and through ecosystems, ultimately transforming into heat. The heat flows from the surface of the Earth through the atmosphere and back into space. (For more details on how energy moves through ecosystems, flip back to Chapter 10.)

REMEMBER

The amount of energy released by the Sun varies by about 0.1 percent every 11 years. This variation isn't enough to explain current trends in global warming.

Variations in orbital movements of the Earth affect the amount of solar radiation that reaches Earth, creating long-term climate cycles called Milankovitch cycles (see Chapter 2). These cycles explain some climate events like the Ice Ages, but they don't explain our current warming trend.

Living in a greenhouse

The impact of solar radiation on Earth's temperature is affected by the gases in our atmosphere. Figure 12-1 shows how some types of gas trap heat as it reflects off the planet's surface:

1. Energy in the form of sunlight arrives at the Earth.
2. Clouds and ice reflect some light away from the Earth.
3. The Earth's surface absorbs most of the light, transforming it into heat.
4. The Earth releases heat back to the atmosphere.
5. Greenhouse gases in the atmosphere absorb heat, re-radiating some of it back toward the surface.
6. Heat slowly transfers from the atmosphere to space.

REMEMBER

Greenhouse gases have a chemical structure that makes them excellent at absorbing heat. Some of the most important greenhouse gases are carbon dioxide, methane, nitrous oxide, and water vapor.

REMEMBER

Because greenhouse gases keep in the heat like the glass on a greenhouse, the warming of the Earth due to these gases is referred to as the *greenhouse effect*.

The greenhouse effect isn't all bad: without any greenhouse gases in our atmosphere, the Earth would be too cold to support life as we know it. Currently, people are concerned about the greenhouse effect because of the rapid rise in global temperatures. An Earth that's too warm isn't going to be good for life as we know it, either.

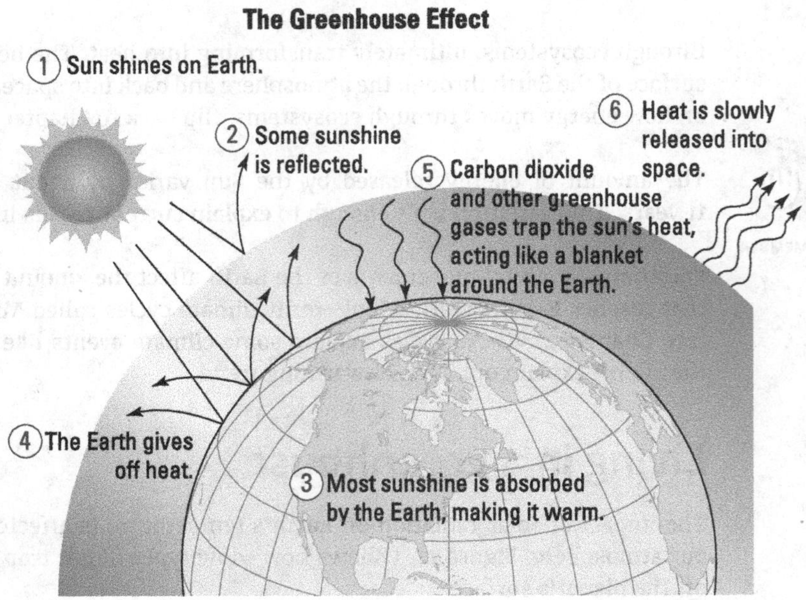

FIGURE 12-1: The greenhouse effect.

Changing the atmosphere

In the 1930s, British engineer Guy Callendar showed that global temperatures were rising. He proposed that the increase might be connected to human influence on the atmosphere (see the sidebar "Citizen Scientist" for more details). Callendar calculated how much carbon dioxide (CO_2) people had added to the atmosphere over the previous 50 years from the burning of fossil fuels and showed how this could explain the global temperature increase.

When scientists today compare global temperatures with atmospheric carbon dioxide concentrations, they see a pattern like the one shown in Figure 12-2:

- **Global average temperatures show yearly variation and a steady upward trend, particularly since the last half of the 20th century.** Temperatures are shown in degrees Fahrenheit.
- **Atmospheric carbon dioxide concentration is steadily increasing.** The rate of increase in CO_2 in parts per million (ppm) correlates with the change in global temperature.

Ever since Callendar first proposed that the burning of fossil fuels might be increasing global temperatures, scientists have continued to collect data to test that idea. The United Nations Environment Programme and the World Meteorological Organization established the International Panel on Climate Change (IPCC) in 1988. The IPCC works with climate scientists from around the world to evaluate the evidence of climate change. Scientists create sophisticated computer programs that use principles of physics and chemistry to simulate the Earth's climate and make predictions about the consequences of climate change. The IPCC assesses and reports on the science of climate change, providing governments with information they can use to establish climate policies.

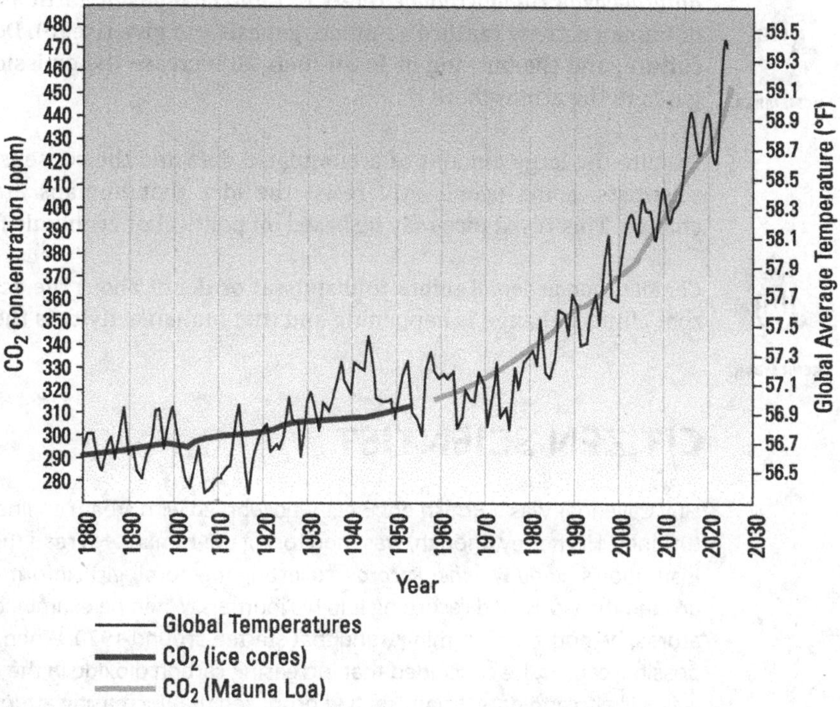

FIGURE 12-2: Global average temperature compared to atmospheric carbon dioxide.

REMEMBER

A *climate model* is a computer simulation of the Earth's climate system, including the atmosphere, ocean, land, and ice.

IPCC scientists compared observations of global temperatures from the mid-1800s to the present with temperature predictions from three different models:

1. A model that predicted global temperatures based solely on natural factors like volcanic activity and variation in solar radiation.
2. A model that predicted global temperatures based solely on human activities like increased concentrations of greenhouse gases.
3. A model that predicted global temperatures based on the combination of natural factors and human activities.

The model that used both natural and human factors was the best match for the actual observations of global temperature changes, and the model based on human activities was also very close, leading IPCC scientists to conclude that the majority of observed global warming since the mid-1800s is due to anthropogenic factors.

REMEMBER

Anthropogenic climate change refers to rapid changes in Earth's climate as a result of human activity (anthro = human, genesis = to give rise to). Deforestation, agriculture, and the burning of fossil fuels all increase the emissions of greenhouse gases to the atmosphere.

Despite the large amount of accumulated data and the consensus among climate scientists, some people still resist the idea that humans are causing climate change. This resistance may be based on political or economic factors.

REMEMBER

Climate change denial refers to dismissal or doubt about the scientific consensus that climate change is happening and that human activity is the primary cause.

CITIZEN SCIENTIST

Guy Callendar was a British engineer who worked with steam engines and turbines. His hobby, though, was monitoring the climate. He read the Smithsonian Institution's *World Weather Records*, collecting meteorological information from around the world and recording it in his journals. When he examined global temperatures, he noticed a warming trend that started around 1920. When he dug into the possible causes, he concluded that increasing carbon dioxide in the atmosphere was a likely candidate. Scientists had proposed that increasing atmospheric carbon dioxide might increase global temperatures, but Callendar was the first to say that it was already happening and that humans were behind it. To support his idea, he did complex calculations (by hand!) to estimate how much carbon dioxide people had generated by the burning of fossil fuels and showed that this could explain the warming trend. He wrote up his research and presented it to the Royal Meteorological Society in 1938.

Callendar's ideas weren't accepted by scientists right away, even though he backed up his claims with solid support. Then, as now, some people just weren't ready to accept that human activity could change the climate. Callendar continued his research hobby throughout his life, inspiring debate and investigation by other scientists. Evidence supporting Callendar's hypothesis continued to accumulate in the second half of the 20th century, eventually leading to consensus among climate scientists that the burning of fossil fuels was the main cause of global warming. This huge shift in thinking about climate was triggered by someone who wasn't a climate scientist, but had a powerful interest and excellent research skills. Today, many scientific organizations, from NASA to the National Park Service, put out calls for citizen scientists to help with research. Some contributions, like transcribing documents for the national archives, can be done from your own home; other projects might involve taking pictures along a trail in a national park. Not all citizen scientists will cause huge changes like Guy Callendar, but, like him, everyone can participate in something that interests them.

Tracking Changes

Fossil and geologic evidence show that Earth's climate has changed periodically over its long history. As discussed in Chapter 2, climate can change from natural causes like volcanoes, ocean currents, Earth's orbital variations, and changes in solar output. One of the biggest differences between the climate change that's happening today and the events of the past is the speed at which things are changing. According to the evidence, the Earth is warming faster today than it has in 10,000 years.

Monitoring the atmosphere

Thanks to a geochemist named Charles Keeling, climate scientists have an almost continuous record of atmospheric carbon dioxide since 1958. Keeling started measuring atmospheric carbon dioxide as part of his research on carbonate rocks in Big Sur, California. This led to an invitation from the U.S. Weather Service to set up an atmospheric monitoring station at a brand-new observatory on the Mauna Loa volcano in Hawai'i. The volcano's remote location makes it a great place to get relatively clean atmospheric samples.

The carbon dioxide measurements from Mauna Loa, called the Keeling curve, are shown on the right side of Figure 12-2. Carbon dioxide is measured in parts per million (ppm), indicating the number of CO_2 molecules within every million molecules of air. The Keeling curve revealed patterns in the variation of atmospheric carbon dioxide:

» Carbon dioxide varies slightly on a daily cycle, with concentrations lower during the day and higher at night, corresponding to the pattern of photosynthesis and cellular respiration. During the day, plants absorb more CO_2 from the atmosphere for photosynthesis than is released by organisms doing cellular respiration. When photosynthesis stops at night, the contribution of cellular respiration to the atmosphere can be measured (for more details on the carbon cycle, head back to Chapter 11).

- Seasonal variations also occur. In the Northern Hemisphere, atmospheric CO_2 rises in winter, reaching its highest concentration at the start of Spring. Then, when leaves emerge and photosynthesis speeds up, atmospheric CO_2 decreases again. This cycle, which some scientists poetically describe as the biosphere breathing, isn't as strong in the Southern Hemisphere.

REMEMBER

Despite daily and seasonal variation, the measurements from Mauna Loa show that atmospheric CO_2 concentration has been climbing steadily since 1958.

Observations from other atmospheric stations around the world support the patterns seen at Mauna Loa.

Inspecting ice cores

Scientists can access information on historical climates from ice cores. Scientists drill into the ice pack in the Antarctic, pulling out long tubes of ice that include bits of history in the form of gas bubbles, pollen, ash, and dust. The surface layer represents the present, but as scientists go deeper, they access time periods going back thousands of years. Scientists analyze the gases in the trapped pockets of ancient air and use pollen, ash, and dust to establish timelines. For example, trapped dust can be compared to dust of known volcanic eruptions to figure out which layers match specific dates. The left side of Figure 12-2 includes data on atmospheric carbon dioxide taken from ice cores.

Scientists can use ice core data to piece together changes in other greenhouse gases (GHG) in addition to carbon dioxide. Figure 12-3 shows the increase in greenhouse gas concentrations in the atmosphere over the last 2,000 years. Things looked pretty steady until just about 200 years ago, when carbon dioxide, methane, and nitrous oxide began to increase rapidly. One of the biggest changes during that time period was that humans entered the Industrial Revolution and began using coal and other fossil fuels to power new industrial processes. Since the Industrial Revolution, the concentration of all three of these greenhouse gases has increased in the atmosphere:

- CO_2 from 280 ppm to 430 ppm (parts per million).
- CH_4 from 715 ppb to 1920 ppb (parts per billion).
- N_2O from 270 ppb to 338 ppb (parts per billion).

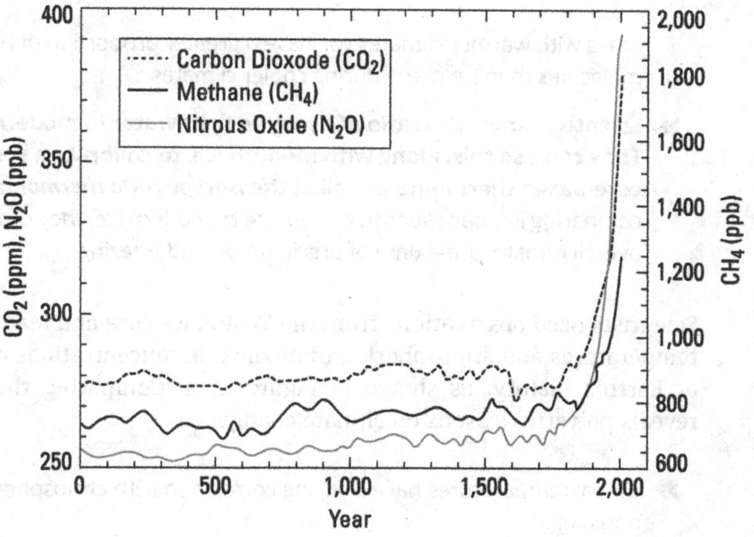

FIGURE 12-3: Increasing greenhouse gases in the atmosphere over the last 2,000 years.

Some ice cores allow us to access information from the very distant past. Scientists studying the Vostok ice core have determined both atmospheric CO_2 and global temperatures stretching back about 400,000 years before the present day. Scientists can determine the temperature of the Earth when ice was deposited by comparing the ratios of types of hydrogen and oxygen atoms in the water molecules that form the ice. It's a little complicated, but here are the basics:

- **Chemically identical atoms of an element can be heavier or lighter, depending upon how many neutrons are in the atom.** Scientists call these different forms of atoms *isotopes*.
  - **The most common isotope of hydrogen doesn't have any neutrons at all and has an atomic mass of 1.0.** A heavier isotope of hydrogen, called deuterium, has one neutron and an atomic mass of 2.
  - **The most common isotope of oxygen has a mass of 16 from its 8 protons and 8 neutrons.** A heavier isotope has two extra neutrons, giving it a mass of 18.
- **All isotopes of an element have the same properties and can join with other atoms to form molecules.** Most water molecules (H_2O) will contain the most common forms of hydrogen and oxygen. But a small fraction of water molecules will contain the heavier isotopes.
- **It takes more energy for heavy water molecules than lighter ones to evaporate from seawater, so it's more likely that heavier water molecules will make it into the atmosphere when the climate is warmer.** After the heavier water molecules make it into the atmosphere, they're more likely than lighter water molecules to fall in precipitation. So, ice made during

- times with warmer climates will have a greater proportion of heavy water molecules than ice made during cooler climates.
- **Scientists know the ratio of heavy to light water in modern seawater. They can use this, along with modern ice, to calibrate a sort of ice core-based thermometer called the *isotope ratio thermometer*.** By comparing isotope ratios in ancient ice to modern ice, they can estimate the overall climate at the time of precipitation and freezing.

Scientists used observations from the Vostok ice core and Mauna Loa to compare temperatures and atmospheric carbon dioxide concentrations over 400,000 years of Earth's history, as shown in Figure 12-4. Comparing these measurements reveals powerful lessons on climate change:

- Global temperatures have a strong correlation with atmospheric carbon dioxide.
- Low drops of the curve represent ice ages, and the high points represent interglacial periods like the current one.
- The rate of change that occurred in the past, even during very rapid ice age transitions, is much lower than today's rate of change visible on the far right of the graph.

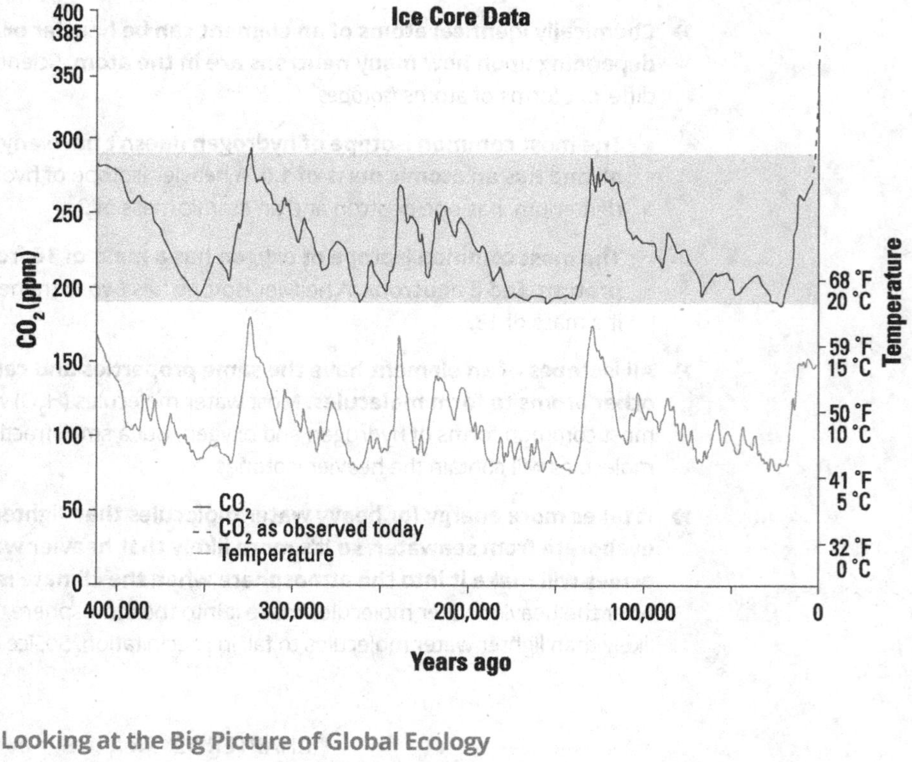

FIGURE 12-4: Carbon dioxide and temperature for the last 400,000 years.

Predicting the Effects of Rapid Global Warming

Earth's climate has changed many times over its long history, as you can see in Figure 12-4. The ice age cycles that occurred about every 100,000 years were periods of glacial expansion followed by shorter interglacial warming periods during which the glaciers receded. As glaciers were expanding and retreating, species moved to locations where they could survive, shifting ecosystems with the ebb and flow of the ice. One reason that species were able to survive these periods of natural climate change was that they didn't happen very quickly on a geological timescale.

REMEMBER

The *Last Glacial Maximum* (LGM) is the time during the most recent ice age when ice sheets were at their largest size, which was approximately 21,000 years ago.

Since the LGM, global average temperatures have gone up about 6 degrees Celsius (11 degrees Fahrenheit). The temperature started warming about 18,000 years ago and continued to increase for about 7,000 years. In contrast, the speed of warming in modern times is more than 10 times faster, with a rise of 1 degree Celsius since the 1970s.

Climate change is happening so quickly that species may not have enough time to migrate or adapt, leading to profound consequences for human health and the environment:

- **Increases in food insecurity:** Climate determines the type and location of human-managed ecosystems, such as agricultural farmlands. Some areas that are currently very good for growing important staple crops like grain may no longer support those crops in the future. Farmworkers and livestock may suffer from heat-related health issues.
- **Reduced quantity and quality of fresh water:** Rising temperatures affect where, when, and how much precipitation falls. Some areas are flooding while other areas experience droughts. Flooding increases the risk for water pollution, reducing water quality. Drought affects water availability. According to the United Nations, terrestrial water storage has steadily declined for the past 20 years, reducing the amount of fresh water that is available for human use.
- **Increased severity of droughts, storms, and floods:** Climate change is already affecting humans around the globe to the point that we have a new category of refugees called *climate refugees*. Climate refugees are people who have had to move from their homes because of climate change.

High temperatures, heavy downpours, floods, and droughts have all been occurring more frequently in the United States over the last 50 years.

- **Changes in ecosystem functions.** Arctic ecosystems are warming twice as fast as the global average, causing ice to melt and sea levels to rise.
  - **Coastal areas will be under severe pressure as increasing temperatures raise sea levels, causing coastline erosion and flooding of low-lying areas.** Changing ocean temperatures may affect the nutrient cycles of the coastal waters, affecting their productivity. People who live near the coast may lose their homes to flooding. Many fish and migratory bird species will lose important breeding grounds.
  - **The stress of warming ocean water contributes to *coral bleaching,* when corals expel their symbiotic algae. Bleached coral isn't dead, but it's more susceptible to disease and death.** Coral reefs provide food, shelter, and breeding grounds for thousands of marine species. Scientists estimate that half of the world's coral reefs have already been lost to climate change.
  - **Increased carbon dioxide is causing *ocean acidification.*** As seawater absorbs carbon, it becomes more acidic and damaging to many organisms, especially those that form coral reefs. We can also expect more severe weather along our coasts, contributing to flooding and coastline erosion.
- **Climate affects soil quality and thus plant growth.** Scientists predict that extreme precipitation caused by climate change will increase soil erosion in many areas around the world. Increased runoff from precipitation can also leach important nutrients from the soil. Changes in water and nutrient availability in soils will lead to changes in vegetation types.
- **Changes in species distribution.** As conditions change, biomes will shift, and organisms may no longer be able to survive in changing conditions (for details on biomes, head back to Chapter 2). Some organisms may be able to migrate or disperse to new areas, but others may face extinction. According to scientists at the University of Southern California, about half of the world's animals are already moving to cooler environments.
  - **As species move into new areas, they may become *invasive species* that harm the environment or species already living there.**
  - **Organisms that carry human diseases may migrate into new areas, increasing risks for human health.** Warmer temperatures and wetter environments favor the survival and reproduction of mosquitoes, allowing mosquitoes that carry viruses like Zika and dengue fever to move into new areas in the United States.

- **Species that can't move or that lack adaptations to survive in new environments may become extinct.** The last five major *mass extinctions*, events where great numbers of species went extinct in a relatively short period of time, are shown in Figure 12-5. Three of the five major mass extinctions were associated with increased greenhouse gases and ocean acidification due to volcanic activity. Many scientists think we are entering a sixth major mass extinction as a result of anthropogenic climate change.

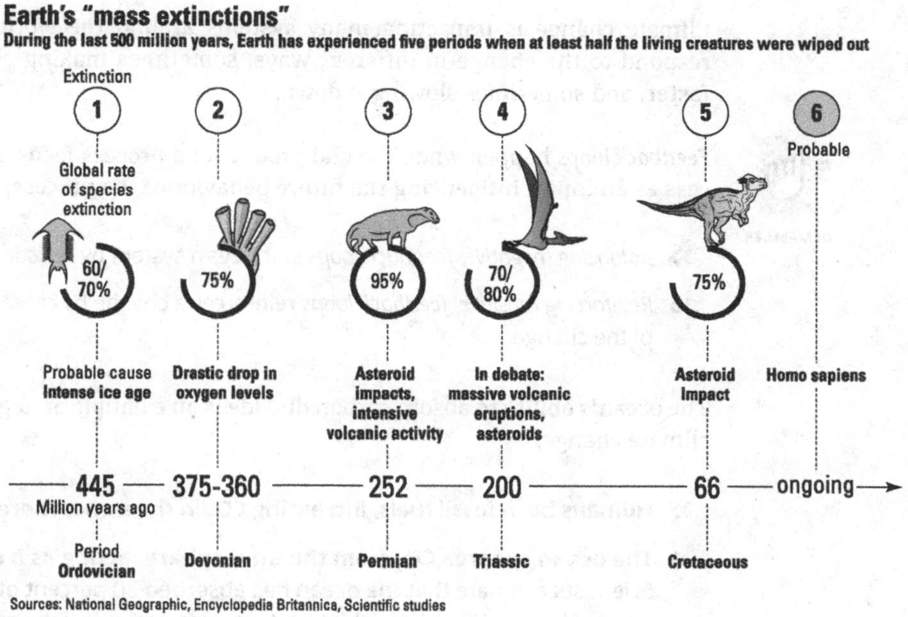

FIGURE 12-5: Mass extinctions.

Regulating Climate Change

Climate change is happening because of human impacts on the carbon cycle (presented in detail in Chapter 11). Carbon that was stored in the Earth as fossil fuels is being released into the atmosphere as carbon dioxide. When forests are cut down to make room for farms or homes, the trees no longer act as carbon sinks that take carbon dioxide out of the air and store it in their bodies as organic carbon. Governments around the world came together to make a plan to slow or stop climate change by reversing these behaviors.

REMEMBER

The *Paris Agreement* is a legally binding international treaty on climate change that was adopted by 195 Parties at the UN Climate Change Conference (COP21) in Paris, France, on December 12, 2015.

The goal of the Paris Agreement is to limit the increase in the global average temperature to less than 2°C above pre-industrial levels. Countries that signed the agreement agreed to create climate change action plans that outlined how they would reduce their greenhouse gas emissions and how they would deal with the consequences of climate change. For more on how well countries are meeting their goals, see the sidebar "The Paris Agreement."

Checks and balances

Climate change is impacting many systems around the world. These systems respond to the change in different ways, sometimes making climate change go faster, and sometimes slowing it down.

REMEMBER

Feedback loops happen when the end product of a process feeds back into the process as an input, influencing the future behavior of the process:

- *Balancing (negative) feedback loops* stabilizes a system by resisting a change.
- *Reinforcing (positive) feedback loops* reinforces a change by causing even more of the change.

The ocean's ability to absorb carbon dioxide is an example of negative feedback on climate change:

- **Humans burn fossil fuels, increasing CO_2 in the atmosphere.**
- **The ocean absorbs CO_2 from the atmosphere, acting as a carbon sink.** Scientists estimate that the ocean has absorbed 30 percent of the carbon released since the Industrial Revolution, greatly slowing the rate of climate change.
  - **The *oceanic biological pump* (shown in Figure 12-6) begins when phytoplankton at the ocean's surface absorb CO_2, assimilating it into organic molecules.** Carbon travels deeper into the ocean as it moves through the food web. When organisms die, they sink into the depths. Some of their carbon may get released back into deep waters as CO_2 by decomposers, while some becomes organic detritus that settles to the bottom of the seafloor. Once covered with sediment, this organic carbon may remain at the bottom of the ocean for very long periods of time.
  - **The *physical biological pump* begins when atmospheric CO_2 dissolves into surface ocean waters, forming bicarbonate ions (HCO_3^-).** As ocean water cools, it becomes more dense, dragging the carbon down into the deep sea. Because bicarbonate dissolves best in cold water, the oceans at the Earth's poles are especially important to the carbon cycle.

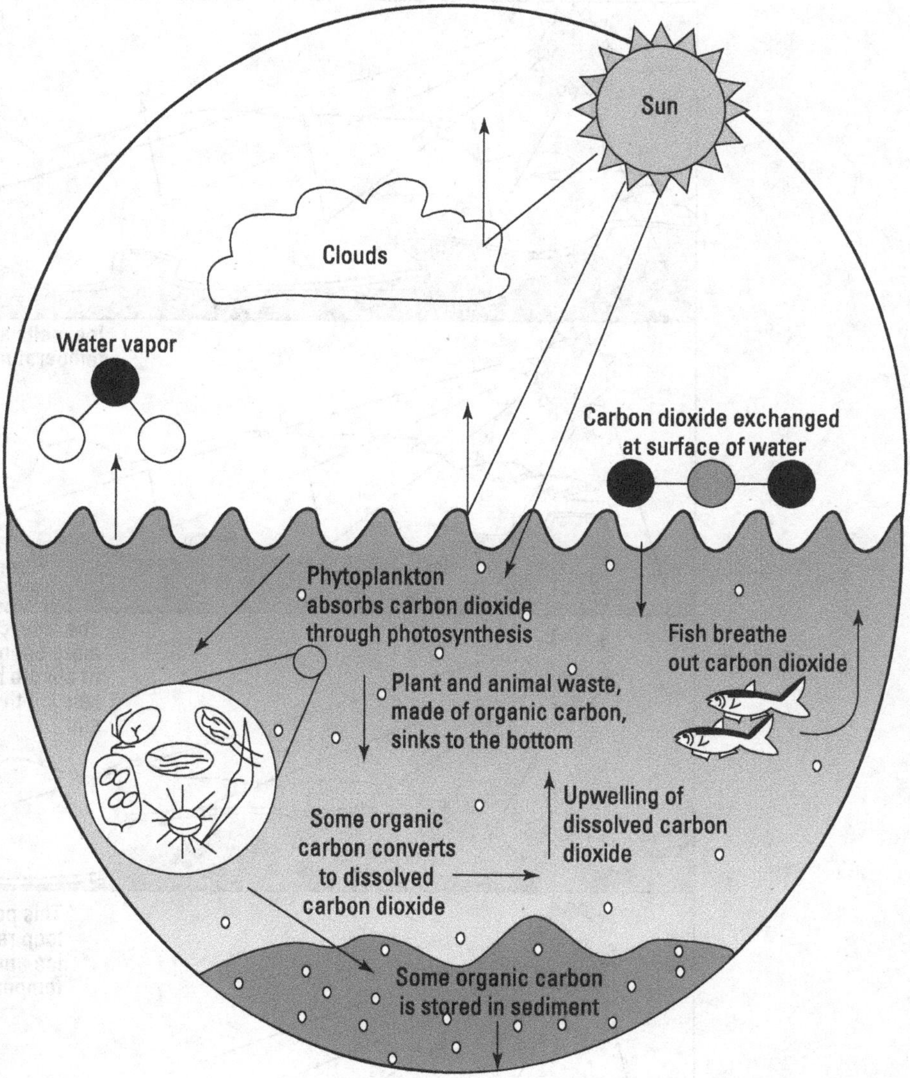

FIGURE 12-6: The relationship between carbon dioxide and the oceans.

The melting of polar ice is an example of positive feedback on climate change. Ice has a high albedo, which means it is light-colored and normally reflects solar radiation out into space (see Figure 12-7). Water and land have a lower albedo and absorb solar radiation:

- Increased global temperatures cause ice to melt.
- Melting ice reveals darker colored land or water.
- The Earth's surface absorbs more solar radiation, increasing global temperatures.

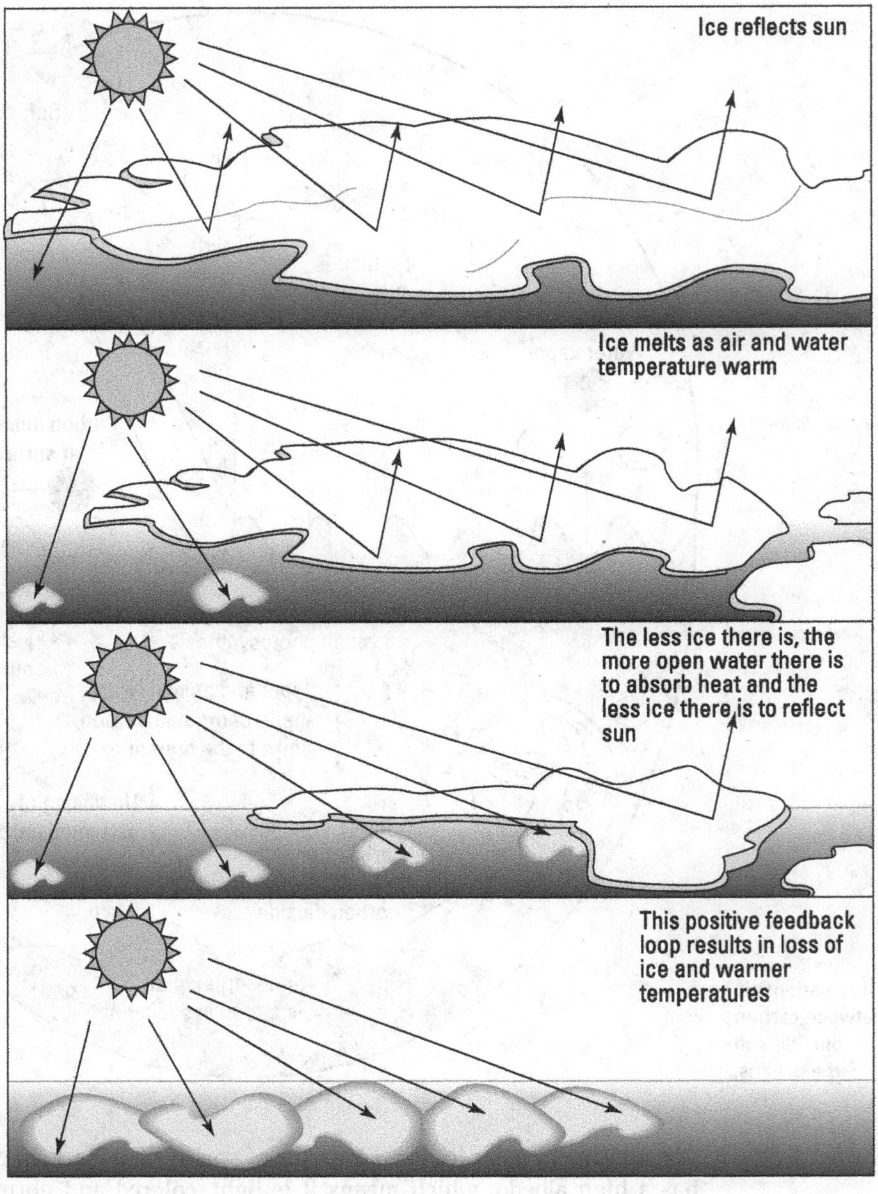

FIGURE 12-7: Ice reflects heat, but water absorbs it.

Tipping points

You may have heard people mentioning tipping points when they were talking about climate change.

REMEMBER

Tipping points are thresholds to dramatic system changes.

One way to think of them is as points of no return or moments when many small changes accumulate to a point that a system collapses and a new organization takes its place.

REMEMBER

Climate tipping points occur when changes in the climate system become irreversible after a warming threshold. For example, if global temperatures cross the threshold for the melting of Antarctic ice sheets, then it will be impossible to stop rising sea levels.

Scientists have identified many climate tipping points, including the melting of large ice sheets in Greenland and West Antarctica, the death of low-latitude coral reefs, and widespread sudden permafrost thaw. Scientists estimate that the warming thresholds for these tipping points range between one and six degrees Celsius above pre-industrial averages and warn that multiple tipping points may be triggered if global temperatures reach an anomaly of 1.5 degrees Celsius. Because of this, some world leaders are saying people need to intensify their efforts to fight global warming. The IPCC stated that to achieve this goal, greenhouse gas emissions must stop increasing by 2025 and decline by 43 percent by 2030.

THE PARIS AGREEMENT

Since signing in 2015, countries have met annually to discuss their efforts and progress toward the goals of the Paris Agreement. These meetings are called "Conference of the Parties" or COP for short. The United Nations also monitors global efforts to reduce carbon emissions through its Intergovernmental Panel on Climate Change (IPCC). In their 2023 report, called the AR6, the IPCC said that humans have the tools necessary to stop the increase in carbon dioxide emissions, but that countries aren't acting fasting enough. The opportunity to stop global temperatures from warming to 1.5 degrees Celsius above pre-industrial levels is rapidly disappearing, and the window to hold warming to 2.0 degrees is very close. In their report, they stress the importance of preserving ecosystems that act as important carbon sinks and encourage nations to develop in ways that are clean and climate-friendly.

The barriers to limiting global warming are mostly financial and political. Although the United States is one of the top three emitters of carbon dioxide, its commitment to the Paris Agreement has not been consistent. President Obama signed the Paris Agreement

(continued)

(continued)

on September 3, 2016, but then President Trump announced in June 2017 that the U.S. would withdraw and began the withdrawal process in November 2020. On February 19, 2021, President Biden officially signed the U.S. back into the agreement. But again, one of the first things President Trump did when he took office the second time was to withdraw the U.S. Neither President Biden nor President-elect Trump attended the Conference of the Parties of the United Nations Climate Change Conference (COP29) in November 2024. Currently, the United States is the only country to have withdrawn from the Paris Agreement.

IN THIS CHAPTER

» Mapping biodiversity

» Exploring factors that influence biodiversity

» Losing biodiversity

Chapter 13 Measuring Biodiversity

Scientists have identified and named more than two million species on Earth, but that's just the tip of the iceberg because so many species have yet to be described. Based on the number of known species and the rate at which species are being described, scientists estimate that the number of species on Earth is closer to nine million. And when scientists factor in the lesser-known groups like microbes and insects, those estimates go up as high as six billion, with microbes making up 70–90 percent of the total. Although scientists don't yet have a firm estimate on the total number, it's clear that humans are just one species out of many. In this chapter, you'll explore what biodiversity is and why it's important, then take a look at the factors that affect it.

Valuing Biodiversity

For as long as people have existed, they've been noticing and recording the species that live around them. Early humans painted animals on the walls of caves 45,000 years ago. Throughout history, indigenous people all over the world verbally passed their knowledge of local plants and animals from one generation to the next. The Ancient Greeks wrote down their ideas about the relationships between different organisms, beginning the Western tradition of identifying, classifying, and cataloging life that was expanded and refined by Linnaeus in 1753. Naturalists like Charles Darwin, Maria Sibylla Merian, and John Muir traveled the world in the 18th and 19th centuries, adding new species to the growing catalog of life on Earth. These naturalists were the predecessors of today's ecologists, who continue to survey and monitor species, seeking to understand the forces that shape the diversity of life forms on the planet.

REMEMBER

Biodiversity refers to the variety of life on Earth at all its levels, from genes to ecosystems.

Scientists measure the diversity of life at different levels:

- **Species richness, or taxonomic diversity:** The number of species may be surveyed or collected from available inventories of regions like countries, parks, or islands. These data can be converted to the number of species per a standard unit of area, then overlaid onto a map to create a picture of biodiversity like the one in Figure 13-1. Using species richness as a measure of biodiversity is a good way to get a quick overview of the biodiversity of an area. Still, it does have some limitations, so ecologists often consider other factors as well:
  - **Species evenness:** The number of individuals representing each species also contributes to the diversity of a region. If two areas have the same number of species but a few species heavily dominate one, while the other has a more even mix, the one with more evenness is actually more diverse. To account for this, ecologists often use measures of species richness and species evenness to calculate a diversity index that represents both factors (see Chapter 8 for details).
  - **Phylogenetic diversity:** Species diversity alone doesn't always capture whether an environment can support different types of organisms. For example, scientists have identified over 350,000 species of beetles. Imagine if you were in an area that had 100 different species, but they were all beetles. Compare that to an area where you have 100 different species, but they're a combination of different plants, mammals, birds, lizards, and fish. The environment with greater *phylogenetic diversity*, or diversity in taxonomic levels, has greater biodiversity.
  - **Endemism:** Some species have a high degree of *endemism*, meaning they are found only in certain locations. When ecologists think about preserving biodiversity, they often take this factor into account. Once a species with high endemism is lost from its particular environment, it's lost from the entire world.
- **Genetic Diversity:** Ecologists also evaluate biodiversity at the genetic level to see what genes are present in an ecosystem. This approach can look at diversity within the species level (for more on genes and alleles, head back to Chapter 3).

» **Ecosystem Diversity:** A third way ecologists can look at biodiversity is through the variety of ecosystems in an area. An area that includes multiple biomes is likely to have higher biodiversity than an area that is homogeneous (for more on biomes, flip to Chapter 2).

REMEMBER

Biodiversity has multiple dimensions, including the richness and abundance of species (taxonomic diversity) and the presence of different evolutionary lineages (phylogenetic diversity).

The naturalists and ecologists of the past had to rely upon direct observation of species to create inventories. Today, ecologists may use DNA technology to supplement or substitute for these observations, as shown in Figure 13-2:

1. **eDNA sampling:** DNA is extracted from environmental samples collected from soil, water, or air. Scientists call this DNA *environmental DNA* or *eDNA.*
2. **eDNA metabarcoding:** Scientists choose a gene to *amplify* from the eDNA. The choice is based upon the group of organisms being assessed; for example, a gene known to be a good marker for animals would be different than that for bacteria.
    a. Amplification uses the *polymerase chain reaction (PCR)* to make many copies of the selected gene in the sample. This results in various copies from different individuals in the target group who left DNA traces in the sample.
    b. *DNA sequencing* is used to read the code in the amplified genes. The code in each amplified gene is represented by colored bars, one color for each of the four nucleotides in DNA. This creates a colorful visual pattern that scientists call a *DNA barcode*. The DNA barcode represents a highly variable portion of a DNA sequence that differs significantly between taxa, enabling the identification of species or strains.
    c. Scientists compare the amplified genes to known sequences stored in computer databases. Species are identified when a sample sequence matches a known sequence. In some cases, no match may be found, which can indicate the discovery of a new species.
3. **Evaluate results:** Information from eDNA metabarcoding can be compared to traditional observations. As shown in Figure 13-2, observations from catch and release allow estimation of the actual abundance of species in the environment (for more details on catch and release, head to Chapter 16). The results from eDNA metabarcoding reveal the relative abundance of DNA from different species in the environmental sample, but that doesn't necessarily match actual species abundance. One advantage of eDNA metabarcoding is that it can reveal small organisms, including microbes, that traditional observational methods might miss.

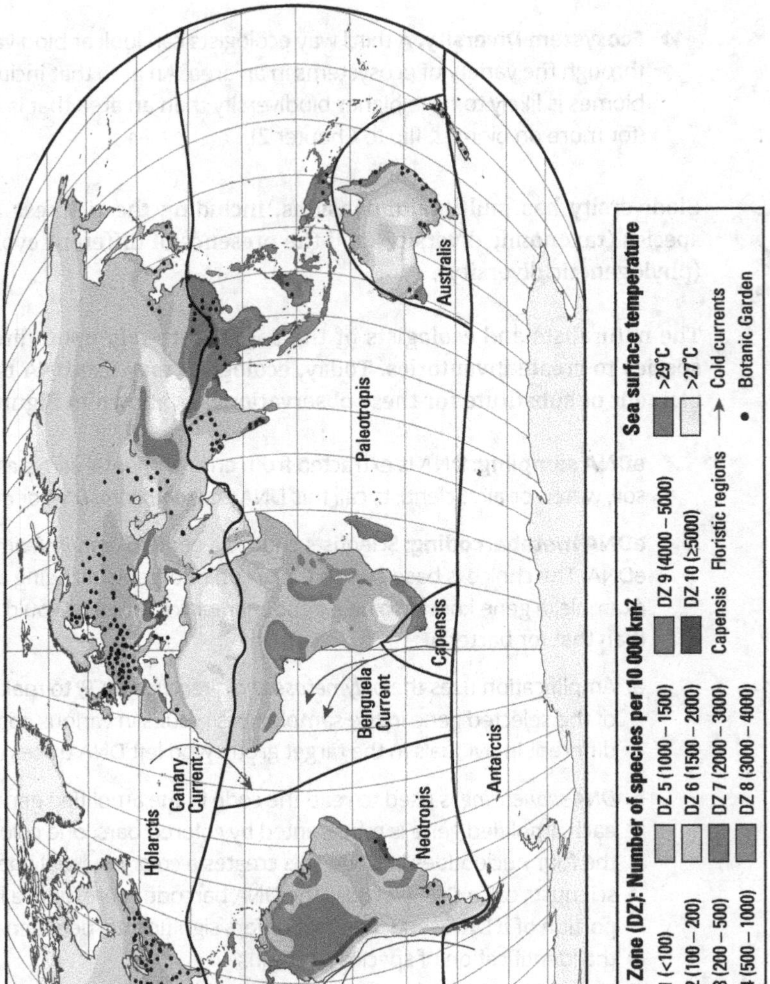

FIGURE 13-1: World map of the species richness of vascular plants.

(Based on Barthlott, Wilhelm, Kier, Gerold, and Mutke, Jens. 1999. Biodiversity - The Uneven Distribution of a Treasure. NNA Reports. 12/1999. 18-28.)

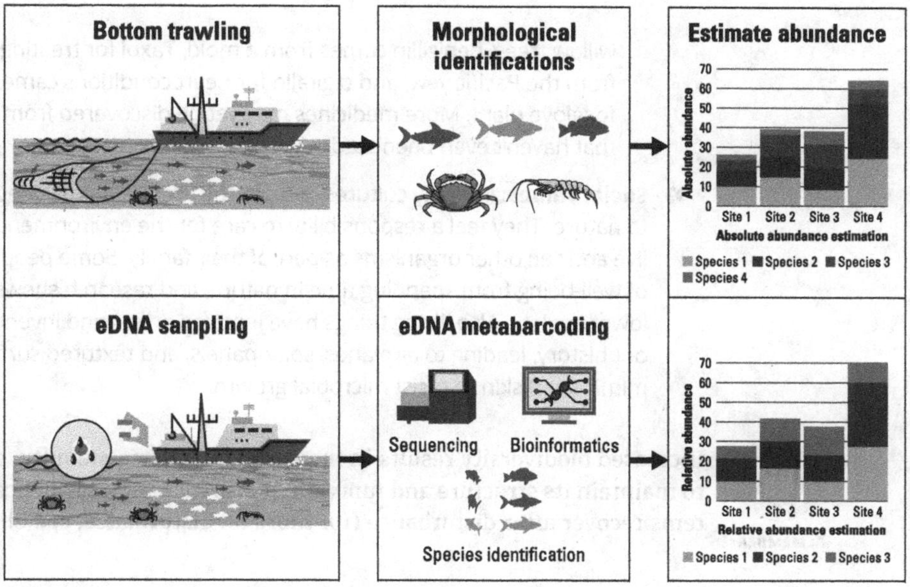

FIGURE 13-2: Comparison of direct observation with eDNA metabarcoding.

REMEMBER

For methods like eDNA metabarcoding to be useful, scientists must first observe and identify species, sample their DNA, and enter reference sequences into databases.

Species inventories provide essential baseline records for scientists who work to conserve biodiversity. Biodiversity has many different types of value, each of which might appeal to different people:

- **Intrinsic value:** Each living thing has value and a right to exist. Some people might say this is a moral way of looking at biodiversity.
- **Utilitarian value:** Organisms have value because of what they provide to others. Utilitarian values for biodiversity include goods, services, and information:
  - **Food:** People today have access to diverse food crops grown all around the world. Many plants grown only in tropical regions, like coffee and chocolate, find their way into the everyday lives of people living in cooler climates. These crops have utilitarian value to the people who work in the agricultural industries that grow, harvest, and process these foods, too.
  - **Ecosystem Services:** Higher biodiversity improves ecosystem functions, resulting in greater plant productivity, soil fertility, water quality, and air quality.
  - **Medicine:** Over half of modern medicines are derived from plants. For example, the compound in aspirin used to be gathered from the bark of

willow trees, penicillin comes from a mold, Taxol for treating cancer came from the Pacific yew, and digitalin for heart conditions came from the foxglove plant. More medicines may yet be discovered from organisms that haven't even been studied.

- **Social values:** In many cultures, people see themselves as deeply connected to nature. They feel a responsibility to care for the environment in which they live and see other organisms as part of their family. Some people get a sense of well-being from spending time in nature, and research shows that it can lower anxiety. Also, living things have inspired artists and inventors throughout history, leading to airplanes, solar panels, and textured surfaces that mimic sharkskin to resist microbial growth.

REMEMBER

Increased biodiversity results in increased ecosystem stability, enabling a system to maintain its structure and function. It also increases resilience, helping ecosystems recover after disturbance (for more on disturbance, see Chapter 9).

Finding Patterns of Diversity

To look for global patterns of biodiversity, ecologists place information on species diversity onto maps like the one in Figure 13-1, which uses shading to indicate ten different diversity zones for vascular plants. The darker the shading, the greater the number of plant species per area in that region.

Vascular plants are one of the best researched groups in terms of biodiversity and are the most important producers in terrestrial ecosystems. Perhaps because of this, the global patterns in biodiversity for vascular plants correlate with those of other species, including insects and land animals:

- **In general, biodiversity increases along latitudes from the poles to the equator.** The light-colored areas near the poles shown earlier in Figure 13-1 represent arctic tundras and boreal conifer forests with low species richness. The dark-shaded areas that have the highest species richness occur in the humid tropics and subtropics. Biodiversity also varies within single latitudes, indicating that local variations also have an influence. For example, areas around the Mediterranean have higher diversity than other parts of Europe at the same latitude. Also, numbers of mammal species increases from east to west in North America. These patterns in biodiversity correlate with changes in temperature and topography.

» **Variation in topographic features of the landscape, especially the presence of mountains, increases regional biodiversity.** All six of the most biodiverse places on Earth occur in mountainous areas in the humid tropics and subtropics.

» **Primary productivity is an important predictor of biodiversity.** Ecosystems with higher levels of primary productivity can typically support larger populations and more complex food webs. In studies of bird biodiversity, net primary productivity was the most important influence on species richness.

» **Increasing temperature and water positively correlate with biodiversity for most groups.** Increasing temperatures generally increase biodiversity, as long as water is also available. In regions with uniform temperatures, like the tropics, the availability of water becomes a more significant influence on biodiversity.

REMEMBER

The *energy-diversity hypothesis* predicts that areas of the Earth that receive more energy will have higher biodiversity.

Areas that receive more energy typically have higher temperatures, which can accelerate metabolism and reproduction. In addition, increases in solar energy mean higher productivity and thus more food. Ecologists sometimes use *potential evapotranspiration (PET)* as an indicator for both the water and energy status of an ecosystem. PET is the amount of water that could be evaporated by the soil and transpired by plants, given the average temperature and humidity of an area. PET values show a positive correlation with species richness of North American vertebrates.

Viewing different scales

Biodiversity can be measured over different areas, from local to regional scales:

» *Local*, or *alpha diversity*, looks at species richness over a small homogeneous habitat.

» *Regional*, or *gamma diversity*, measures species richness over all habitats in an area, as long as organisms are free to move between the habitats.

Local and regional diversity would be the same if each species were present in every habitat within a region. However, if some habitats have unique species, then regional diversity would be greater than local diversity.

Ecologists measure the difference between habitats within a region as *beta diversity*. A higher beta diversity means more species differences between habitats.

Ecologists calculate beta diversity by comparing similarities and differences in species between habitats. For example, Sørensen's similarity index calculates the proportion of common species between two habitats as:

$$\text{Sørensen similarity} = C/[(S_1 + S_2)/2]$$

where C is the number of species in common between two habitats divided by the average number of species in those habitats (S_1+S_2/2). If all species were the same between the two habitats, then Sørensen similarity would equal one, and there would be no beta diversity. If all species are different, then C equals zero, Sørensen similarity is zero, and beta diversity is high.

REMEMBER

Looking at beta diversity helps ecologists identify which species can move between habitats, helping to preserve biodiversity. This is important for understanding how biodiversity might be affected by large-scale environmental changes like global warming.

Considering relationships

The ability of species within a region to move between habitats can help stabilize biodiversity. The organisms within a regional *species pool* have the potential to replace individuals that are lost from local habitats. Species can potentially move into habitats that fall within their fundamental niche. Whether or not a species becomes established will also depend upon its interactions with other species, including predators and competitors. (For more on these interactions and the concept of fundamental niche, see Chapter 7.)

REMEMBER

Species sorting refers to the process that selects whether species in a regional pool become part of a community. This process depends upon adaptations of the species and its interactions with other organisms.

In areas with high levels of competition, each species is most likely to survive in areas where they are the best competitor. In other words, species will be limited to their *realized niche* and likely to be found in fewer communities. Habitats within a region will be more unique, and beta diversity will be high.

The importance of interactions like competition in determining the realized niche can often be seen on islands. If an island has fewer species on it than a nearby mainland, then competition on the island is reduced. When a species on an island is compared to the same species on the mainland, the population on the island has often spread out into more habitats. In other words, the population on the island occupies a broader niche.

REMEMBER

Ecological release is the expansion of a species from its realized niche to its fundamental niche in a habitat with lower competition.

Maintaining Equilibrium

When the naturalist E.O. Wilson was studying ants on islands in the South Pacific, he wondered why certain species of ants were present on some islands and not others, and why some islands had more species overall. His research revealed the importance of island size and the distance from islands to the mainland. Together with Robert H. MacArthur, Wilson synthesized his ideas into the understanding that biodiversity represents a dynamic balance between processes that increase diversity, like immigration or speciation, and processes that reduce diversity, like emigration and extinction. A model of these ideas is shown in Figure 13-3.

REMEMBER

The *equilibrium theory of island biogeography* states that the number of species on an island results from a balance between regional processes of immigration and local processes of extinction.

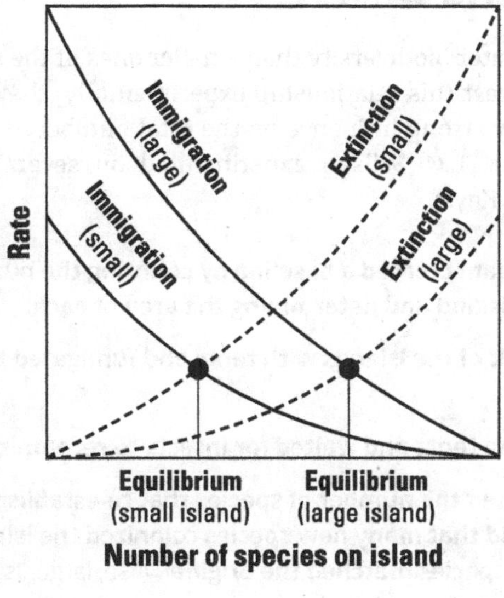

FIGURE 13-3: The equilibrium theory of island biogeography.

Islands are often useful systems for studying ecology, acting as natural laboratories that are slightly less complex than larger ecosystems. The lessons learned from island studies can be applied to larger ecosystems, too. For example, conservation biologists apply the principles learned from island biogeography when they create plans to preserve biodiversity within disturbed environments. As shown in

Figure 13-4, fragmented patches of habitat can be seen as analogous to islands floating in a sea of human development. The creation of wildlife corridors between patches allows immigration from one "island" to another, helping to stabilize biodiversity.

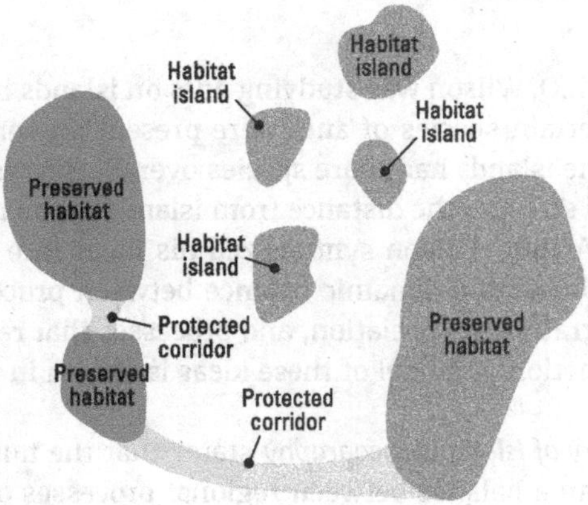

FIGURE 13-4: Connecting habitat islands with corridors.

The bigger the better

Larger areas have greater biodiversity than smaller ones of the same type of habitat. When ecologists test this relationship experimentally, they find that the size of a habitat has a very strong influence on the total number of species that it can support. For example, E.O. Wilson experimented on several small mangrove islands in the Florida Keys:

1. **Wilson and his team created a baseline by counting the number of insect species on each island and determining the area of each.**
2. **They covered half of the islands with tents and fumigated them to kill all the insects.**
3. **They removed the tents and waited for insects to recolonize the islands.**
4. **When they counted the number of species that re-established on each island, they found that many new species colonized the islands, but the total number of species matched the original.** Also, larger islands supported more species than small islands.

Ecologists have conducted variations of this experiment in many different environments with similar results. Figure 13-3 illustrates that small and large islands reach different equilibrium points with regards to biodiversity.

REMEMBER

The *species-area* relationship says that larger geographic areas generally have greater biodiversity than smaller areas.

Larger areas typically have a great diversity of habitats, which supports a greater diversity of species.

Another factor in determining the biodiversity equilibrium is the extinction rate. Figure 13-3 shows that small islands have a higher rate of species extinction than do large islands.

Finding a target

One reason that larger islands may support higher biodiversity is that they may attract more immigrants.

REMEMBER

The *target effect* is the hypothesis that larger islands are more likely to be encountered by chance by immigrating species.

Ecologists tested the target effect by monitoring migratory birds that visit islands off the shores of Ireland and Great Britain. Larger islands had higher visitation rates than small islands. Figure 13-3 shows that the rate of immigration is higher for large islands than for small islands. Both immigration rates decrease over time because the chance of a new species discovering an island decreases when more species are present. In other words, over time, it's more likely that repeat species will discover the island.

Living far away

Another factor that affects biodiversity on islands is the distance between the island and the mainland. The mainland is the source of potential immigrants to the island, so the greater the distance, the less likely it is that new species will travel from the mainland to the island.

Ecologists tested this idea by comparing the number of arthropods that live in desert shrubs. The arthropods don't survive well outside the shrubs, so the shrubs act like islands in an ocean of desert. One particularly large patch of shrubs acted as a mainland source of arthropods for the scattered shrubs. When the ecologists compared the number of species present in each shrub, they found that shrubs farther from the mainland had fewer species.

REMEMBER

The *distance effect* predicts that the equilibrium number of species on an island decreases as the distance between the island and a source of species increases.

Losing Biodiversity

As the human population grows and pressures on ecosystems intensify, more species will be driven to extinction, reducing biodiversity. By studying ecosystems and the factors that impact biodiversity, ecologists hope to identify ways to predict which systems are most at risk of collapse and help mitigate these impacts.

The most significant factors that put pressure on ecosystems and lead to loss of biodiversity can be remembered by the acronym HIPPO, which was first proposed by the American naturalist, E.O. Wilson, in his book *The Creation*:

"The decline of Earth's biodiversity is an unintended consequence of multiple factors that have been enhanced by human activity. They can be summarized by the acronym HIPPO, with the order of the letters corresponding to their rank in destructiveness."

Habitat loss

The most destructive impact on biodiversity is from *habitat loss*, the conversion of natural areas for human use. Eight billion humans need homes and places to grow their food, plus we build factories to produce goods, shops to sell those goods, and many more conveniences of modern life.

Humans spread out from cities into rural areas, taking over chunks of land bit by bit. Human modifications to the environment have many potential negative impacts on biodiversity:

- **Fragmentation of ecosystems:** When large wild areas are split into smaller remnants, some species can't survive. For example, top predators often need large areas for their hunting grounds. For these species, several small, separated bits of an ecosystem don't equal the same amount of space in one unbroken piece. Sometimes, people can help alleviate the problem by creating wildlife corridors, like the ones shown in Figure 13-4, that allow animals to move from one smaller ecosystem to another.
- **Monocultures:** Humans reduce genetic and ecosystem diversity by planting large crops of one type of plant. Over half of the prairie ecosystems in the United States have been converted into croplands, reducing biodiversity and leading to soil erosion.

» **Habitat destruction:** When people consume natural resources, such as timber or minerals, they destroy natural habitats. Cutting trees from a forest doesn't just remove the trees and the habitat for the birds; it also removes habitat for insects, refugia for animals, and completely alters the microbial environment. In some areas, species and environments are being lost before scientists even have a chance to study what was there.

Invasive species

As people travel and move to new areas, they bring other species with them, whether it's a favorite food plant from home, a companion animal, or even an accidental microbial hitchhiker. These *introduced species,* or non-native species, can sometimes become invasive, outcompeting native species and displacing them from an ecosystem. When this happens, it can cause a ripple effect through an ecosystem, leading to a loss of biodiversity (for more on how this can happen, flip back to Chapter 9).

One of the most devastating invasions occurred when rats were introduced to New Zealand. Before human colonization, no mammals, except for a few species of bats, inhabited the islands. The niches on the islands were filled with unique species, with an emphasis on fantastic birds. Giant moa grew up to 12 feet tall, stretching their necks like giraffes to graze on the tops of tall plants. Flightless green parrots called kākāpō hid from predators during the day, then glided down from the trees at night. Some New Zealanders refer to themselves as Kiwis, after funny little pear-shaped birds the size of a chicken with long, thin beaks. The arrival of humans drastically changed the unique ecosystem of New Zealand, driving the giant moa and other large birds to extinction, and putting many other species at risk.

Much of the impact on New Zealand's biodiversity occurred because of the introduction of rats. The Maori brought kiore, or Pacific rats, as food, but the rats multiplied rapidly, expanding through the environment where they preyed on the eggs of ground-dwelling birds and reptiles. When the British arrived, they brought Norway rats and ship rats as stowaways. The impact of rats and other introduced species caused the extinction of about 25 percent of New Zealand's native bird species and pushed many more to the edge.

New Zealanders have taken a strong stance against further extinctions and are working to eradicate rats and other mammalian predators. They're using fencing, trapping, and poison to eradicate the mammals, one area at a time. A few of the smaller islands are now predator-free, providing refugia for endangered, endemic species like the kākāpō.

Pollution

Eight billion people produce a lot of waste. When this waste enters ecosystems, it causes environmental changes that can stress or even kill organisms:

- **Oil spills** kill marine life, devastate ecosystems, and require years for recovery. One of the most recent devastating spills was the Deepwater Horizon spill in the Gulf of Mexico in 2010, which killed many species, including turtles, dolphins, and birds. Oil and chemical dispersants remain in the Gulf today and continue to negatively impact local fisheries and marshlands.
- **Greenhouse gases** enter the atmosphere, trapping heat and causing global warming (see Chapter 12 for details). Carbon dioxide dissolves into the ocean, where it contributes to ocean acidification. Increasing temperatures and ocean acidification are impacting species' survival.
- **Agricultural waste**, like pesticides, may control agricultural pests, but they also kill other species. Fertilizers used to enhance crop growth can spread into local waterways, leading to eutrophication and the death of species (for more details on eutrophication, head back to Chapter 10).
- **Industrial waste**, like sulfur and chemical pollutants, damages ecosystems, leading to losses in biodiversity. Sulfur from the burning of coal enters the atmosphere, where it becomes sulfuric acid. (For more, see Chapter 11.) The sulfuric acid falls to Earth as part of acid rain, damaging forests and creating toxic conditions in aquatic ecosystems.
- **Plastic** is an amazing building material that's used to manufacture everything from packaging to life-saving medical equipment. Unfortunately, plastics don't break down naturally, and that is a huge environmental problem. Dead whales and sea turtles have been found with stomachs full of plastic bags and other garbage. Seals and other marine mammals get tangled up in plastic fishing nets. Areas of our oceans are full of trash, like the Great Pacific Garbage Patch, an area of floating plastic that's twice the size of the state of Texas and located halfway between Hawaii and California in the Pacific Ocean.

Population growth

Eight billion humans already occupy most habitats and use a lot of resources, and our population is still growing (see Chapter 6 for more on the predictions for human population growth). Many people around the world live *unsustainable lifestyles*, meaning they consume resources faster than the biosphere can replenish them. For example, people in the United States are among the top consumers of resources like energy and fresh water, and also among the top producers of greenhouse gases.

In some parts of the world, including East Africa, the central East coast of S. America, and Central America, concentrated human populations occur very close to biodiversity hotspots (see Figure 13-5, later in this chapter). As these human populations continue to grow, they may increase the pressure on biodiversity hotspots and cause the loss of species. Tropical rainforests are one of the most diverse ecosystems in the world, but they are under increased threat of human invasion all the time. In the Amazon, about 17 percent of the forest has been lost in the past 50 years, mostly due to clearing for cattle ranching.

Overconsumption

Humans are also decreasing biodiversity directly by *overconsumption* of certain species, which occurs when we harvest, use, or exploit species faster than their ability to recover:

» **Hunting and fishing:** People are more skilled than ever at locating and killing other species, which can lead to overhunting and overfishing. If we hunt and fish responsibly, food species are a renewable resource, but this requires that people don't kill species faster than they can reproduce. Several factors have contributed to past failures to harvest responsibly:

- **Accurate assessment of wild populations:** Sometimes, it's hard to determine the reproductive rate or the size of a stock population. Wildlife biologists attempt to count and estimate migrating populations of fish, for example, but it's not an exact science.
- **Profit motive:** There's constant pressure between profit and conservation. Biologists may set a harvest number based on their best estimate of what's required to conserve a species. Still, these numbers are often challenged and compromised by people with profit motivation.
- **High-tech tracking:** People have gotten so good at tracking and finding organisms that prey have fewer places to hide. It used to be that fish could hide around rock outcrops, leaving some fish to reproduce in the next season. But equipment like sonar fish-finders makes it possible to locate fish in formerly protected areas.
- **Collateral damage:** Some fishing and harvesting methods harm other species or the ecosystem. Species such as turtles, dolphins, and marine birds often get entangled in fishing nets, becoming *bycatch* that fishermen don't want or aren't allowed to keep. These animals are usually discarded and often die. In some parts of the world, people use explosives to stun fish, making them easy to harvest in mass quantities. Unfortunately, this method also destroys the ecosystem.

» **The tragedy of the commons:** This phrase refers to the probability that people will degrade a shared resource if they act in their own self-interest. The basic idea is this:

- **In any industry that makes a profit by harvesting a natural resource, the profit goes to those who harvest the fastest.**
- **If any individual chooses to harvest slowly, perhaps because they want to save some of the resource, someone else will swoop in, take everything, and use up the resource anyway.** Many fisheries around the world have been fished to the point of collapse because of this type of behavior.

» **Illegal animal trade:** Another tragic example of humans reducing biodiversity is through illegal trade of endangered animal parts and the capture of animals for the exotic pet industry. You're probably aware that elephants are hunted for their ivory tusks, rhinos for their horns, and sharks for their fins. Clouded leopards are killed for their pelts, and parrots are captured for the pet industry, many dying from the horrible conditions during smuggling.

REMEMBER

The root cause of most biodiversity loss today is human population growth and the resulting overconsumption of resources.

Sounding the Alarm

The combined impact of the human activities summarized by HIPPO is reducing biodiversity not just by lowering species richness and phylogenetic diversity, but also by reducing the number of individuals within species. (For the implications of these losses in insects, see the sidebar "A Quiet Apocalypse.") A beautiful silver porpoise called the Vaquita may have just ten individuals left in the world. The kākāpō parrots in New Zealand number just 236 adults. Both of these species are considered *critically endangered* by the International Union for Conservation of Nature, meaning they're at extreme risk for extinction in the wild.

A QUIET APOCALYPSE

When you think of endangered species, what comes to mind? Polar bears standing precariously on melting ice, big cats like clouded leopards padding through the jungle, or maybe a local population of whales that's struggling to survive? Maybe you've heard that honeybees are in trouble, suffering from something called *colony collapse disorder* that seems to cause worker bees to simply disappear.

While the troubles of some species make headlines, others are dying quietly by the masses all around us. German scientists sounded the first alarm in 2017 when they reported a 76 percent decline in the insect population over a period of 27 years. Scientists in Puerto Rico followed with a report that their insect populations had declined by approximately 60 percent since the 1970s. More reports followed, telling tales of falling population numbers of mayflies, butterflies, beetles, and more. Although species weren't necessarily going extinct, their numbers were plummeting to the point where they might no longer be able to fulfill their typical roles. Scientists started talking about *functional extinction* and thinking about how this could affect ecosystems.

Insects are often unsung heroes of ecosystems. Even the less popular mosquitoes and flies provide essential services. Insects are important food for birds, fish, reptiles, and amphibians, and the drop in insect numbers is paralleled by declines in the numbers of animals that rely on insects for food. Bees, flies, wasps, and beetles are important pollinators that are essential to three-fourths of human crops. Insects are part of the clean-up crew that deals with waste and dead material so that nutrients get recycled back into the environment. Without enough insects, ecosystem collapse is a very real possibility. If humans survive, they will face a future of low biodiversity, including a lack of food species.

Some scientists argue that the situation isn't as bad as the early reports suggest, but others say that we need to take action now to prevent the insect apocalypse from occurring. Scientists who are working to save honeybees report that the bees are suffering from a combination of "the four P's": parasites (*Varroa* mites), pathogens (bacteria, viruses, and fungi), poor nutrition, and pesticides. Entomologists who study other types of insects tell the same story: Insects aren't suffering from just one thing; they're suffering from everything all at once.

The World Economic Forum recommends five things you can do to help support insect populations.

- Grow plants, especially those that benefit pollinators.
- Provide structures in your garden that act as shelter or watering stations.
- Stop using pesticides.
- Stop buying food grown with pesticides.
- Reduce light pollution by turning off outside lights at night.

Humans are transforming the Earth and reducing habitat for many species. Insects are small, though, so a little patch of land or even a flowerbox on a balcony can become an oasis that provides food and shelter for a small but important neighbor.

Focusing on hotspots

One way to try to preserve biodiversity is to focus on areas of the world that have high levels of biodiversity and are also at great risk from human activity. These *biodiversity hotspots*, shown in Figure 13-5, make up only about 2.5 percent of the Earth's surface, but they are home to more than half of the world's plants and about 43 percent of land vertebrates. Many of the plant species found in hotspots are endemic. To qualify as a hotspot, an area must have lost at least 70 percent of its native vegetation.

In some areas, local communities are partnering with governments and non-profit agencies to try to protect hotspots. New Zealanders have launched an aggressive campaign to eliminate invasive mammals in order to protect their endemic species. In the Himalayas, local people are partnering with conservation organizations to plant millions of trees. In Peru, indigenous communities monitor their forests with satellite imagery and work to reduce tree loss. And in Indonesia, the government has designated more than 24 million hectares as marine conservation areas.

REMEMBER

The identification of biodiversity hotspots enables focusing of conservation resources on areas where they have the potential for maximum effect.

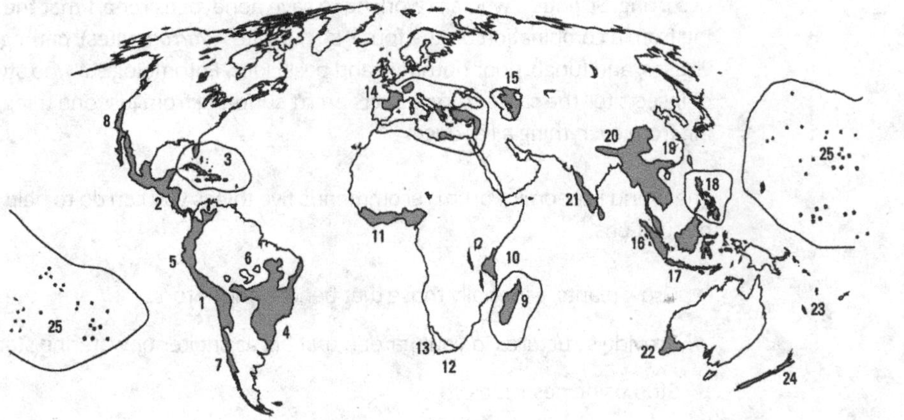

1. Tropical Andes, 2. Mesoamerica, 3. Caribbean, 4. Brazil's Atlantic Forests, 5. Chico/Darien/Western Equador, 6. Brazil's Cerrado, 7. Central Chile, 8. California Floristic Province, 9. Madagascar, 10. Eastern Arc & Coastal Forests of Tanzania/Kenya, 11. West African Forests, 12. Cape Floristic Province, 13. Succulent Karoo, 14. Mediterranean Basin, 15. Caucasus, 16. Sundland, 17. Wallacea, 18. Phillippines, 19. Indo-Burma, 20. South-Central China, 21. Western Ghats/Sri Lanka, 22. Southwest Australia, 23. New Caledonia, 24. New Zealand, 25. Polynesia/Micronesia.

FIGURE 13-5: Biodiversity hotspots.

Mass extinction

REMEMBER

Extinction is when all members of a species have died.

Extinction is a normal part of life on Earth. Based on the fossil record alone, it's obvious that species come and go. Scientists calculate that the normal, or *baseline rate of extinction* is between 0.1 and one species per 10,000 species per 100 years. Now and then, though, the rate of extinction increases and many species go extinct around the same time, geologically speaking.

REMEMBER

A *mass extinction* occurs when species vanish much faster than expected. This is usually defined as about 75 percent of the world's species being lost in a short period of geological time — less than 2.8 million years.

Five major mass extinction events have already occurred during the history of the Earth. Most were triggered by dramatic atmospheric changes caused by volcanic eruptions, meteor impacts, or the rapid expansion of land plants. You're probably most familiar with the mass extinction that killed off 78 percent of all species, including most of the dinosaurs, at the end of the Cretaceous Period about 66 million years ago. The impact of an asteroid in the Yucatan Peninsula is thought to have added to environmental stress by increasing carbon dioxide in the atmosphere and causing rapid temperature shifts.

Many scientists think that a sixth mass extinction is currently underway due to human impacts on the environment. Some scientists estimate that current extinction rates are between 100 and 1,000 times higher than the background, resulting in the extinction of as many as three species per hour or 20,000 each year. As human activity puts stress on ecosystems, populations are dwindling, mostly because organisms can't find enough food or mates to successfully reproduce.

Endangered species sound the alarm for human impacts on biodiversity. When a species becomes endangered, it usually means that its entire habitat has been degraded. Taking steps to save an endangered species doesn't just help that one species; it helps its entire ecosystem. (For one way we can try to save endangered species, see the sidebar "Protecting Our Neighbors.")

Some factors make it more likely that a species will become endangered due to human activity:

- Organisms with a slow reproduction rate, like sharks, may not be able to compensate for human overharvesting.
- Apex predators need a wide geographic area for hunting, yet natural habitats are shrinking as humans convert land for their use.
- Likewise, large animals need lots of food, so they need large areas for foraging.

- Organisms that are specialists for certain habitats are more likely to decline because they can't switch habitats easily.
- Species that already have low numbers are less likely to recover if some members are killed.
- Species that are commercially valuable to humans are more likely to be overexploited.
- Species that are considered a threat to livestock, like wolves, may be hunted to local extinction.

PROTECTING OUR NEIGHBORS

In some ways, people are just doing what all species do: taking the resources we need and multiplying as fast as we can. The difference between the human species and most others seems to be that we can understand the consequences of our actions in a way that's perhaps unique in all the world.

Many human cultures live by a principle of reciprocity, the belief that people should give back to nature in balance with what they take. Some even see plants and animals as members of their nonhuman family, worthy of care and respect. The Haudenosaunee Confederacy looks to the future when making decisions because they believe they must respect the world in which they live, as they are borrowing it from future generations.

Many human cultures don't center values like these as part of their culture, which has created a legacy of environmental change and habitat loss for many nonhuman species. The Environmental Movement of the 1970s helped make many people aware of the growing repercussions of human activity. In the United States, this led to the passage of the Endangered Species Act (ESA) in 1973. The ESA makes it illegal to kill, trap, uproot (plants), modify the habitat of, or engage in commerce of an endangered species or its parts. To designate a species as endangered or threatened, the Fish and Wildlife Service or the National Oceanic and Atmospheric Administration must list the species, designate the habitat where the species is found as critical habitat, and develop a recovery plan to help the species survive and thrive.

Sometimes, people argue against these types of laws, primarily because these laws get in the way of industry and profit. People don't always see the value of protecting one species, especially if this makes it harder for them to work and feed their families. It can be hard to see that these laws aren't really about one particular species as much as they are about an ecosystem in which that species lives. If a species is endangered, it's a signal that humans have impacted that species' ecosystem to a very large degree. And that means we haven't impacted just one species, but potentially the other species that live there too.

IN THIS CHAPTER

» Preserving species and ecosystems

» Building functioning landscapes

» Restoring damaged ecosystems

Chapter 14

Saving Our Future

Ecology is all about observing organisms interacting in their environment to build an understanding of how the biosphere functions. The next step for human societies is to decide what to do with this knowledge. Many ecologists work in fields that apply the lessons of ecology to solve problems that result from human activity. In this chapter, you will explore the practical application of ecology to conserve biodiversity, manage landscapes, and even repair damaged ecosystems.

Thinking about Land Use

Every species affects the land in some way, whether it's a prairie dog tunneling to create a habitat or a predatory bird that removes grazers from its hunting area. In some ways, people are no different than other species in this regard. We modify the landscape to create our homes and take resources from our ecosystems to sustain ourselves. The difference between people and other species is the sheer magnitude of land we're converting to our specific needs.

Figure 14-1 illustrates how human population growth really took off in response to the Industrial Revolution. Perhaps in response to the changes happening around them in the United States during the 19th century, people began to express concern for natural ecosystems. Some, like John Muir, advocated for the rights of other species and the preservation of natural ecosystems. Even among those who saw ecosystems as something for humans to use, people like Gifford Pinchot argued for responsible use that benefited the most people in the long run. (For more details about early conservationists in the United States, see the sidebar "Speaking for the Trees.")

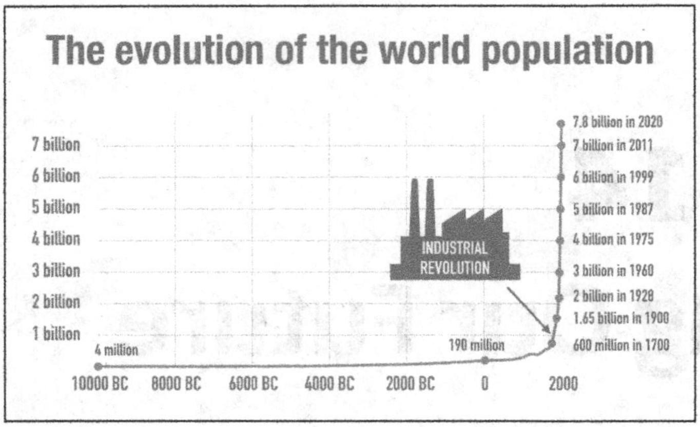

FIGURE 14-1: The impact of the Industrial Revolution on human population growth.

Dimitrios/Adobe Stock Photos

The way people see their relationship to the land and other species is the underlying foundation for the decisions that governments make on a daily basis. Three philosophies dominate these decisions in much of the world:

- *Conservationists* advocate moderate use of resources from ecosystems such that consumption or extraction doesn't damage the ability of the resource to regenerate.
- *Preservationists* say that some ecosystems should be set aside and not used for resources at all.
- *Strictly economic use policies* place short-term human gain, such as profit, over all other considerations, leading to radical exploitation of resources in a way that damages or destroys their regenerative capacity.

SPEAKING FOR THE TREES

The science of conservation biology centers around an ethical principle that biodiversity should be preserved. In other words, conservation biologists apply their ecological knowledge around this goal because it's the right thing to do. The reason *why* it's the right thing is a matter of individual perspective, which can be seen in the differing philosophies of some of the early conservationists in the United States.

Transcendentalists of the early 1800s trusted in the spirit of each individual, believing that people could connect with God through nature. Saving biodiversity, then, was about preserving the natural temple of God on Earth. These ideas about the spiritual value of nature were promoted by Ralph Waldo Emerson (1803-1882) and Henry David Thoreau (1817-1862).

Like the Transcendentalists, John Muir (1838-1914) believed in the power of nature to benefit the human spirit. He guided President Theodore Roosevelt on a three-day wilderness trip through the Yosemite area, helping to convince Roosevelt to consolidate state and federal lands into Yosemite National Park. He also founded the Sierra Club, a nonprofit that still works today to protect the rights of people "to explore, enjoy, and protect the wild places of the earth."

A new philosophy that emphasized practical rather than spiritual reasons became the foundation of the mission statement for many government agencies. Gifford Pinchot (1865-1946), who became the first head of the US Forest Service, appreciated nature but saw it as a resource to be used and managed so that it did the "greatest good for the greatest number for the longest time." In this view, nature is a collection of individual natural resources that need to be managed and controlled by government agencies.

At the same time that the idea of sustainable land use was developing, the science of ecology was contributing to a more holistic view of ecosystems. Aldo Leopold (1887-1948) began his career as a forester, working according to the responsible use philosophies established by Pinchot. As Leopold learned more about the complexity of ecosystems, however, he began to see that organisms didn't fit neatly into categories of natural resources. To preserve nature for the benefit of all, he saw the necessity of focusing on the entire system rather than its individual parts. In his landmark book, *A Sand County Almanac*, he redefined the role of humans in ecosystems from managers to members, writing, "In short, a land ethic changes the role of *Homo sapiens* from the conqueror of the land community to plain member and citizen of it. It implies respect for his fellow members and also respect for the community as such."

Saving Populations

Many societies developed policies and management strategies that attempted to balance human needs with conservation of resources. Governments created agencies that specialized in the management of wildlife, forests, fisheries, and agriculture. Scientists monitor populations and make recommendations for harvesting limits. Hunters and fishermen must get permits and limit their catch to certain numbers and seasons. Despite these efforts, however, biodiversity continues to decline. (For more on biodiversity, turn back to Chapter 13.)

In response to declining biodiversity, a new branch of science began to grow. With roots stretching back to the ideas of people like Muir and Pinchot, but also taking in scientific developments of the 19th and 20th centuries, the science of conservation biology established itself as a new field in the 1980s.

REMEMBER

Conservation biology is centered around protecting biodiversity. It integrates multiple scientific disciplines, including ecology, evolutionary biology, and genetics, along with economics and sociology, to help craft policies that will prevent the extinction of species and preserve biodiversity at all levels.

Conservation biology has three main objectives:

- To document Earth's biodiversity (see Chapter 13 for more on biodiversity).
- To investigate how human activity is impacting species, communities, and ecosystems.
- To develop practical approaches to solve problems, including the extinction of species and loss of genetic variation, with the ultimate goal of protecting biological communities and their associated ecosystem functions.

Threatened and endangered species

The International Union for Conservation of Nature (IUCN) is an international organization with members of governmental and non-governmental agencies from over 170 countries. For over 75 years, the IUCN has promoted cooperation between conservation biologists and societies to protect nature. Beginning in 1964, they began to compile a comprehensive database of the conservation status of species known as the *Red List of Threatened Species*. This list is used by people all around the world to set conservation priorities and guide management decisions. IUCN experts evaluate species based on five factors:

1. **Population size reduction:** The percentage of the population lost over a period of 10 years or three generations, whichever is longer.
2. **Geographic range and evidence of reduction or fragmentation:** This is measured as the *extent of occurrence*, which is the area around all known or projected occurrences of a species, or the *area of occurrence,* which is the area occupied by a species at any one time.
3. **Small population size and decline:** This is based on the number of mature individuals and evidence of population decline.
4. **Very small or restricted population:** This is based on the number of mature individuals and evidence that the population is at risk of becoming critically endangered or extinct in a short period of time.

5. **Probability of extinction in the wild:** This is based on statistical calculations and modeling.

The IUCN places species into categories based on their evaluation. Three core categories, *threatened*, *endangered*, and *critically endangered*, represent species that are threatened with extinction. Currently, 172,600 species have been evaluated and categorized, with more than 48,600 species threatened with extinction. These species include 44 percent of reef-building corals, 41 percent of amphibians, 38 percent of trees, 38 percent of sharks and rays, 26 percent of mammals, 26 percent of freshwater fishes, and 11.5 percent of birds.

Many countries produce their own versions of a red list to document the status of local species. In the United States, at-risk species are identified and protected under the Endangered Species Act (ESA), which was passed in 1973. To designate a species as endangered or threatened, the Fish and Wildlife Service (FWS) or National Oceanic and Atmospheric Administration Fisheries (NOAA Fisheries) must first add the species to the federal list of endangered and threatened wildlife and plants. Like the IUCN, the federal agencies in the USA base their evaluation on the numbers of individuals, geographic area, rate of decline, and risk of extinction.

REMEMBER

According to the ESA, an *endangered species* is "in danger of extinction throughout all or a significant portion of its range." A *threatened species* is "likely to become endangered in the foreseeable future throughout all or a significant portion of its range."

Once invoked to protect a species, the ESA has broad impacts:

- **Protection for the species:** The ESA makes it illegal to kill, trap, or uproot endangered species. Scientists must develop a recovery plan to help the species survive and multiply until the population reaches a healthy size and can be delisted. Scientists monitor the species throughout recovery and make adjustments to the plan as needed.
- **Protection for the habitat:** Scientists designate the habitat where the species is found as *critical habitat*. This habitat is legally protected from modification, which lends protection not just to listed species but also to other species that live in the same habitat.
- **Protection from poaching:** The ESA makes it illegal to engage in the commerce of an endangered species or its parts. Because the USA is part of an international treaty called the Convention on International Trade in Endangered Species of Wild Fauna and Flora (CITES), legal protections under the ESA extend into other countries as well.

REMEMBER

Endangered species sound the alarm for human impacts on biodiversity. When a species becomes endangered, it usually means that its entire habitat has been degraded. Taking steps to save an endangered species doesn't just help that one species; it helps its entire ecosystem.

Analyzing population viability

Ecologists use computer models to conduct *population viability analyses* that evaluate whether species are in danger of extinction. They program the models with information on the species, including life history characteristics (Chapter 5), demographics, and population dispersal (Chapter 6), and may also include information on landscape features. Scientists run these models to test how long a population is likely to persist under various scenarios.

In general, conservation biologists believe that species become more vulnerable to extinction when their populations become small. One question that they frequently try to answer is how small is too small.

REMEMBER

The *minimum viable population size* (MVP) is the smallest number of individuals necessary to prevent a population from going extinct.

Many policymakers want an MVP that they can use when developing management strategies. Unfortunately, many ecologists agree that population dynamics are too fluid in the face of environmental variables to calculate a single "magic number" that fits for a species. In addition, conservation biologists worry that managing populations around an absolute minimum will ultimately doom the species in the face of environmental change.

Based on studies on animal breeding and genetic drift in fruit flies, ecologists in the 1980s proposed the 50/500 rule of thumb:

- **Fifty is the number of breeding individuals needed in the population over the short term to prevent inbreeding.** When populations get small, they lose genetic variability among individuals. Inbreeding can also result in the spread of deleterious alleles throughout a population.
- **Five hundred is the number of breeding individuals needed over the long term to maintain overall genetic variability.** When genetic variability is lost, species may lose the ability to evolve in response to environmental change, increasing their risk of extinction.

The 50/500 rule is based on the number of breeding individuals. When ecologists use these numbers to estimate overall MVP, they propose total population sizes that are much larger than 500 individuals, especially when they take uncertainty into account.

Establishing breeding programs

When efforts to protect species in the wild fail, conservationists sometimes resort to removing members of an endangered species to a protected space where they can breed and recover their numbers.

REMEMBER

Ex-situ (*ex*=out, *situ*=site) *conservation* is the maintenance and breeding of endangered plants and animals under controlled conditions.

You may encounter breeding programs for endangered animals when you visit zoos, aquaria, gardens, or plant nurseries. Ideally, a breeding program provides temporary protection until an endangered species can eventually be released back into the wild. In some cases where no habitat is available, some species become functionally extinct in the wild and exist as remnants in ex-situ facilities.

Animals in captivity typically live longer than their wild counterparts. With protection, good nutrition, and even genetic assistance, populations can recover to a point where they can be released back into the wild. For example, several species have been successfully reintroduced into National Parks in the USA:

- **The California condor once flew the skies of North America from British Columbia to Baja California.** By 1982, however, only 22 of these birds remained, victims of human activities including the use of lead ammunition that poisons the birds' food sources. The remaining birds were captured and taken to a breeding facility, then re-released into Pinnacles National Park. Today, about 210 birds exist in the wild, and 180 remain in captivity.
- **Black-footed ferret populations declined dramatically as humans targeted and killed their prairie dog prey.** In 1987, the 18 remaining black-footed ferrets were taken to a breeding facility for recovery. The National Park Service began reintroducing ferrets into the wild, first to Badlands National Park in 1994, then Wind Cave National Park in 2007, followed by other locations. Today, about 1,000 of this species exist in the wild.
- **The nēnē is the last surviving indigenous goose on the Hawaiian Islands, and they are in trouble.** A combination of human land use changes and the introduction of species like cats, dogs, and mongoose that prey on their eggs reduced the nēnē in Hawai'i Volcanoes National Park from a population in the thousands to just 50 individuals in the 1940s. Captive breeding, reintroduction, and careful monitoring have improved population numbers, but the threats remain.

Captive breeding programs can save species from extinction, but they have several disadvantages. For one thing, they're expensive. In addition, species kept in captivity for long periods may change in response to their new conditions and be less adapted for success in the wild. Finally, reintroduction can be very traumatic for species resulting in high casualties and no guarantee of success.

Working with Landscapes

Landscape ecology focuses on the impact of spatial variation on ecological issues. In other words, it's ecology with an emphasis on the impact of variations like habitat size or *landscape elements*, including lakes, different vegetation types, and even disturbances.

REMEMBER

Landscape ecology is the study of the interactions between the temporal and spatial aspects of a landscape and its ecosystems.

As people transform land for their own use, they increase the heterogeneity, or differences, in the landscape. Landscape ecology studies how landscape diversity, including both natural and modified habitats, impacts ecosystem function. Because of this focus, landscape ecology is especially important for ecological and urban planning that integrates human activity sustainably within the natural world.

Defining landscapes

A *landscape* is a heterogeneous area composed of different elements to form a *mosaic*. The elements may differ in what they're composed of, like different types of vegetation, or in how they're arranged:

- *Landscape composition* refers to the kinds of elements that make up the *patches* in a landscape.
- *Landscape structure* refers to the physical configuration of the elements, including size and arrangement within the landscape.

TIP

You can think of a landscape like a patchwork quilt with squares of different colors and sizes. The differently colored patches might represent different vegetation types, like a green square for a forest, a brown square for a farm, a blue square for a lake, and a gray square for a city.

Landscapes are studied at different levels of *scale*, or size, depending upon the question being asked. For example, in a study that looks at a large geographic area, a forest may be treated as a single, homogeneous unit within a larger landscape. A different study that focuses on just the forest itself might see trees, shrubs, and clearings as different units within the smaller landscape.

REMEMBER

The *patch-mosaic model* refers to the view of landscapes as a mosaic of diverse patch types.

Landscape ecologists use numerous methods and tools to collect, visualize, analyze, and model data over space and time:

- **Field methods** involve ecologists traveling to a landscape to make direct observations or conduct experimental studies. Sometimes, ecologists use tools like drones or other local sensors to collect data. Field studies are particularly good for collecting data on a fine scale.
- **Remote sensing** detects and monitors the physical characteristics of an area by measuring its reflected and emitted radiation at a distance, usually via aircraft or satellites. Remote sensing is especially useful for collecting data on large areas.
- **Geographic Information Systems (GIS)** use computing power to attach data to locations on maps. Multiple types of information can be placed in separate layers on the map. For example, one layer might have information from field studies, while another might have remote sensing data. Still other layers might have information on socioeconomic factors or biodiversity. Layering different types of information on the same map allows people to visualize the entire system and gain unique insights into how different factors contribute to its function. This approach is a powerful way to bring different stakeholders in a region together around the same map to discuss how to solve problems and plan for the future.

Interacting with landscapes

Landscape patches exist within a background ecosystem type called a *matrix*. In landscapes dominated by humans, like the one shown in Figure 14-2A, the matrix is often developed land while patches are remnants of the original natural ecosystems.

Although the patches in a landscape may appear distinct, they're connected by the flow of materials, energy, and organisms that move between them. The movement of animals between patches can be affected by the matrix between the patches. If the matrix doesn't favor the survival of a species, it may only be able to move if suitable corridors exist between patches, as shown in Figure 14-2B. Organisms can move more freely when the matrix is suitable, as in Figure 14-2C.

REMEMBER

Limitations on movement due to unfavorable matrix conditions have direct effects on the animals trying to move between patches and can have indirect effects on other species. For example, restrictions on movement may prevent an animal from pollinating another species or helping disperse its seeds. If the animal is a predator, its prey population might increase in patches it can't reach. If it's prey, its predator might not find enough food to survive.

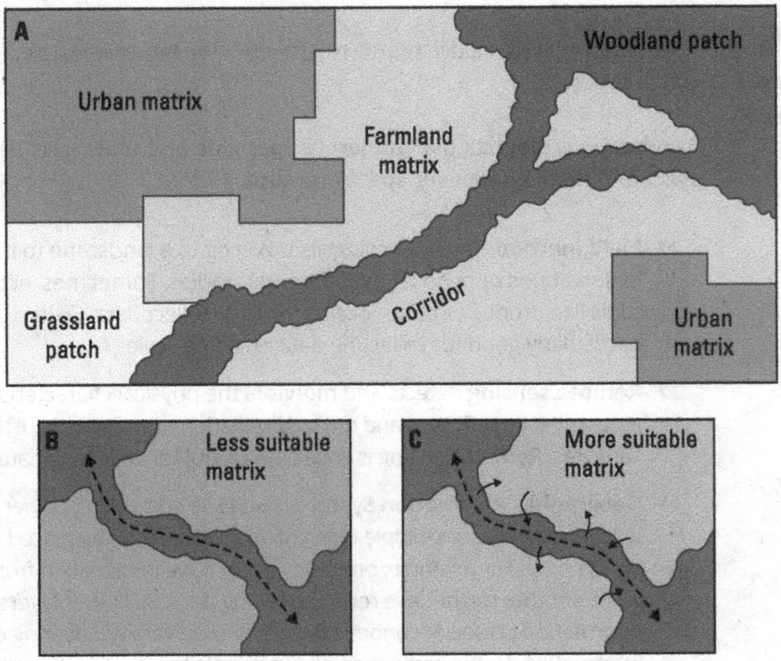

FIGURE 14-2: Patches, matrix, and corridors in a landscape.

Edge effects refer to the abiotic and biotic differences associated with patch boundaries. Figure 14-3 shows how habitat fragmentation causes a higher ratio of edge to core area in fragments. Edges are exposed to the area outside the patch, leading to differences in abiotic factors like temperature, wind, and light. These differences in the environment affect biotic factors like decomposition and community structure. Species that are successful in edges may be more prevalent in smaller fragments, while species that require the conditions present in the core may be lost from small fragments. As a result, smaller habitat fragments generally lack top predators.

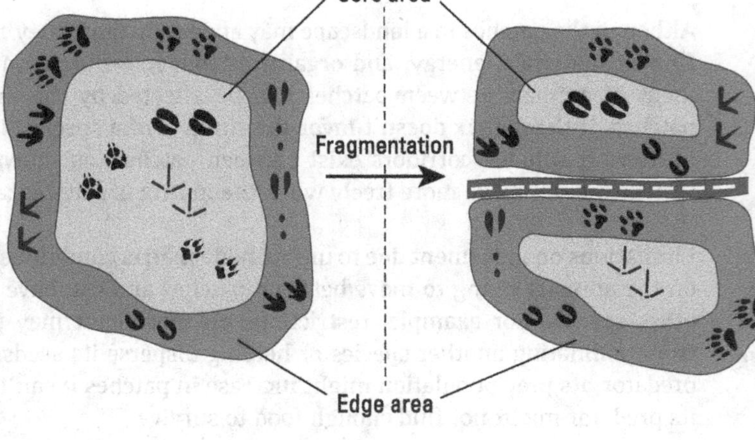

FIGURE 14-3: Edge effects on habitat fragments.

Disturbance and landscape characteristics interact with each other in several ways (for details on disturbance, go to Chapter 9):

- **Disturbance can leave lasting impressions on landscapes, creating unique patches called *landscape legacies*.** The type of disturbance may affect the shape of the patch, which in turn can affect its characteristics through edge effects. A tornado, for example, might remove vegetation in a long strip, leading to a patch with a high ratio of edge to core.
- **Recovery after a disturbance can be influenced by the landscape.** For example, the recovery of a forest after a disturbance that creates several small patches might be quicker than one that opens a large patch of the same total area. Seeds from the surrounding forest will have less distance to travel into the small patches, speeding up recolonization.
- **Landscape characteristics can influence disturbance progress.** Accumulations of flammable material in forests can provide ladders for fire to reach the crowns of trees, allowing it to spread farther and faster through the canopy.

Establishing protected areas

One way that conservation biologists try to preserve biodiversity is to locate and design protected areas called *biological reserves*. Figure 14-4 illustrates how principles from landscape ecology can improve the design of reserves:

a. **Bigger is better:** Large reserves can hold more species at equilibrium and will have lower extinction rates.

b. **A single habitat is better than fragments of the same total area:** Larger habitats hold more species and have less impact from edge effects.

c. **If the reserve is fragmented, keep fragments close together:** When fragments are closer, it's easier for species to pass between them.

d. **Connected fragments are better than disconnected ones:** Connections are better. Habitat corridors allow species to move more freely between patches, enabling immigration and gene flow.

e. **Circular shapes are better than linear ones:** A circular shape minimizes dispersal distances within the reserve. In a linear reserve, organisms might have to travel longer distances to move from one end to another.

f. **Habitat fragments with *buffer zones* improve the effectiveness of protected areas:** A buffer zone allows a transition between the outside matrix and the core area of the fragment. If a reserve is embedded in a matrix of human-impacted land, the buffer zone might be an area that is similar to the core environment but allows managed human use, like logging and hunting. People can still get the resources they need without going into the core, and the core is protected by having a compatible transition zone.

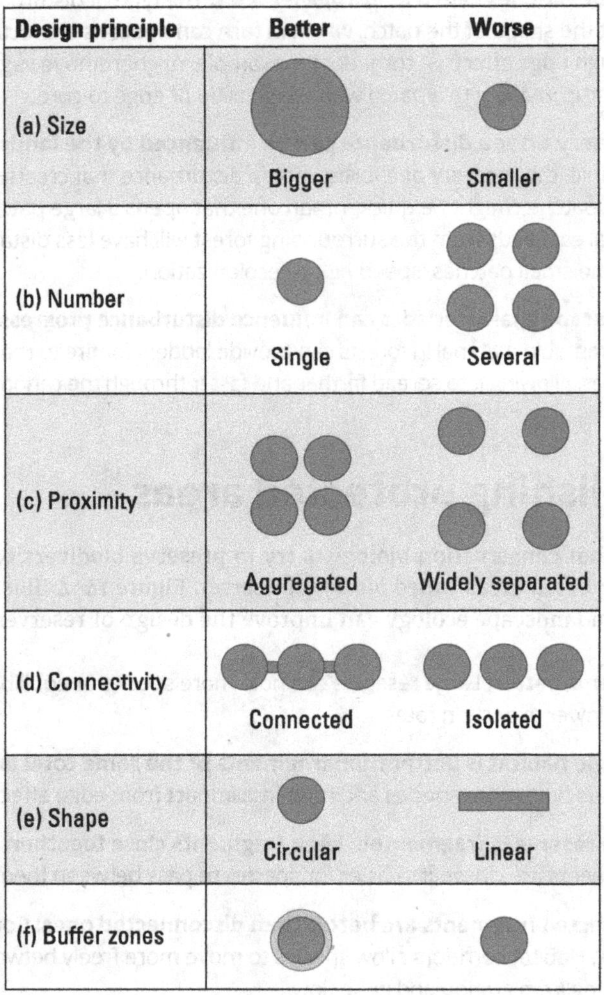

FIGURE 14-4: Recommended spatial configurations for core natural areas.

Restoring Damaged Lands

Biological communities can recover from many types of disturbances through ecological succession (see Chapter 9 for details). In parts of the world, however, the natural rate of recovery can't keep up with the rate of degradation by human disturbance. In this situation, restoration ecologists may apply the lessons of ecology to give nature a helping hand.

REMEMBER

Restoration ecology seeks to restore the health, integrity, and sustainability of ecosystems that have been degraded by human activities.

Restoration ecologists try to start or speed up the recovery of degraded ecosystems. By identifying and manipulating the processes that limit the speed of recovery, they can shorten the time it takes for a community to rebound from disturbance. In some cases, this involves rebuilding the structure of a site through revegetation.

Restoration projects differ in their objectives and methods. Some restoration projects might try to reestablish ecosystems composed of native species. In contrast, others focus on restoring ecosystem function, like pollination or erosion control, with species best suited for the goal. Two strategies that are often used to repair and improve ecosystem function are bioremediation and bioaugmentation.

Cleaning up pollution with bioremediation

REMEMBER

Bioremediation uses living organisms, like prokaryotes, fungi, or plants, to detoxify polluted ecosystems.

One of the most well-known examples of bioremediation is the use of microbes to clean up oil spills, as shown in Figure 14-5:

1. After an oil spill, physical clean-up methods remove as much of the oil as possible.
2. Microorganisms that can digest oil are sprayed over the area. Some of these microbes digest oil naturally, but genetically engineered species may also be used.
3. Nutrients that support the growth of the microbes are added to the environment to support their growth. In some cases, the microbes have been genetically engineered to require a certain nutrient.
4. The microbes degrade the oil, helping to clean the environment.
5. Scientists monitor the environment. More nutrients may be added to encourage microbial growth.
6. Once the oil is gone, the population of oil-eating microbes will decline. If they were engineered to require a nutrient, then scientists would stop adding that nutrient, so their population would decline faster.

In addition to oil-degrading microbes, scientists have used microbes and plants that take heavy metals out of the environment to clean up areas polluted by mining and the development of nuclear weapons. Scientists are currently working to identify or engineer microbes that can break down plastics.

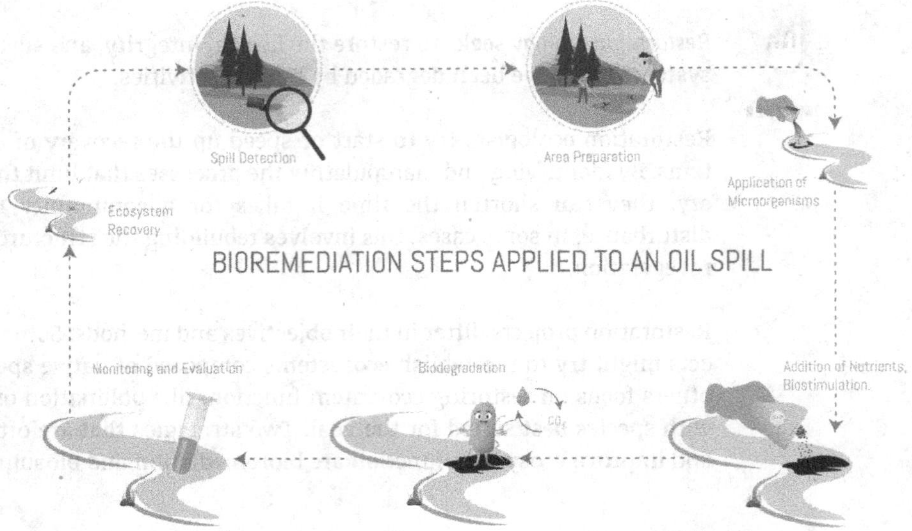

FIGURE 14-5: Bioremediation.

María Pilar Martínez/Adobe Stock Photos

Speeding up recovery with bioaugmentation

REMEMBER

Bioaugmentation uses organisms to add essential materials to a degraded ecosystem to speed its rate of recovery.

Plant growth in an area with degraded soils may be limited by nitrogen availability or a lack of mycorrhizal fungi. To counter this, ecologists may introduce plants that associate with nitrogen-fixing bacteria (for details, see Chapter 11) or add mycorrhizal fungi to the soil. By improving the soil, ecologists help other plants colonize the area, speeding up the recovery process.

Searching for Sustainability

REMEMBER

The United Nations defines sustainability as meeting the needs of the present without compromising the ability of future generations to meet their own needs.

Currently, human beings aren't living sustainably. Scientists estimate that we need 1.8 planet Earths to sustain our population sustainably at its current levels of consumption. We only have one.

The lessons from ecology and conservation biology can help us forge new ways of managing our populations so that we tread more lightly on Earth's ecosystems.

The current increases in global temperature (Chapter 12) and losses in biodiversity (Chapter 13) make it pretty clear we need a new vision.

Traditional ecological knowledge

Conservation biology grew out of Western science, but it's not the only tradition of conservation in the world. Many cultures have long practices of living in balance with their natural environment. These traditions rely on intimate knowledge of their local environment and traditions that foster living in balance with the natural world.

REMEMBER

Traditional ecological knowledge is a cumulative body of knowledge, practice, and belief, handed down through generations by cultural transmission about the relationship between organisms and their environment.

Traditional ecological knowledge is based on generations of practice in a particular ecosystem and is typically passed down as part of an oral tradition. It results from detailed observations and deep knowledge of local ecosystems.

Some cultures, like the Haudenosaunee Confederacy, include principles of sustainability in their teachings. They hold a core value called the *Seventh Generation*, believing that they must look seven generations into the future when making decisions and live in the world as if they are borrowing it from future generations.

Other cultures live by a principle of *reciprocity*, the belief that people should give back to nature in balance with what they take. Some even see plants and animals as members of their nonhuman family, worthy of care and respect.

Systems thinking

People often focus narrowly on the problem right in front of them, failing to see the big picture. Scientists who are struggling against all odds to save a species from extinction may focus on trying to increase population numbers without trying to address the factors that caused the population decline in the first place. People who make land-use decisions often operate on the local level without considering the impacts on the larger region. Individuals may prioritize short-term needs and economic gains, overlooking the long-term impacts of decisions. Governments pass laws and make decisions for the land they control, often without regard for the cumulative impacts on the biosphere.

If we continue these patterns, the World Economic Forum estimates that by 2050, 90% of land on our planet could become degraded.

International cooperation on systemic land-use approaches is needed if we want to avoid this fate. Tools like GIS can help scientists see the biosphere as a single system that needs to be managed as a whole.

REMEMBER

Systems thinking recognizes that the structure of a system drives its behavior. It is a way of analyzing problems that considers all the parts of a system and how they interact with each other.

Systems thinking can help move us toward a sustainable future because it requires us to look at the consequences of our actions on the system itself. As an example, every product that people create requires energy and materials to make and produces waste as a result. If we consider where the energy and materials come from and where the waste goes, then we can see the real impact of that product on the entire system. And, when we see ourselves as connected to and dependent on that system, we can better evaluate whether that product is sustainable.

Human population growth is starting to slow down, and nations are trying to work cooperatively to slow down some of the biggest impacts of climate change, but people may not be changing fast enough. Systems thinking can help with better long-term decision-making, but bringing people together to figure out solutions to global problems may take too much time.

Naturalist E.O. Wilson proposed a solution for what people can do while they're figuring out what to do. He proposed that humanity should set aside half of the world's land and seas as biological reserves to preserve biodiversity and ensure a functioning biosphere.

REMEMBER

The Half-Earth Project works to promote collective action directed at preserving biodiversity. They provide tools like the *Half-Earth Map* and *Species Protection Index* that allow people to see how well countries are protecting biodiversity and meeting their conservation targets.

5 The Part of Tens

IN THIS PART . . .

Make sense of the natural world

Apply ecology to improve your relationship with Mother Nature

IN THIS CHAPTER

» Designing ecological research

» Sampling populations

» Modeling and mapping environmental data

Chapter 15
Ten Methods for Ecological Research

Ecologists seek to understand the natural world, so it's probably no surprise to hear they spend a lot of time outdoors making observations and testing their ideas. The classic image of an ecologist is someone out in nature, peering intently at something and scribbling in a notebook. Sometimes, ecologists are content to observe; other times, they take experiments out into the field or try to re-create nature back in the lab. In this chapter, you'll get an introduction to some of the ways that ecologists make sense of the natural world.

Field Studies

The heart of the science of ecology is the field study, also called field research or just "field work."

REMEMBER

Field studies involve the collection of data in real-world settings rather than in the laboratory.

FIELD STUDY EXAMPLE

After over 300 reindeer were killed on a mountain plateau in Norway by a lightning strike, ecologists got special permission to let the carcasses decompose in place. The scientists wanted to understand the ecological changes that accompany death and decomposition in the Arctic tundra, so they set up cameras and observed the changing landscape over time. Juices from the rotting carcasses initially killed nearby plant life, but then seeds left behind in the feces of scavenging birds and foxes sprouted to give rise to new plants.

Field studies allow ecologists to study things you can't bring into a lab, like a forest, or study things over long periods of time, like the recovery of ecosystems on volcanoes after eruptions. Ecologists take their equipment out into the environment they want to study, hiking through forests and meadows, wading through rivers, and even scuba diving in the ocean to make their observations. If you like the outdoors, ecology can be very fun. It's also a lot of work with plenty of physical activity in all kinds of weather.

Field studies may be strictly observational. Observational studies often involve identifying species and counting the number of individuals in each. Ecologists can't always count every organism, so they use sampling methods like the ones described later in this chapter.

Field studies often include controlled experiments. Ecologists can set up plots in an ecosystem, some that receive an experimental treatment and some that do not.

REMEMBER

Controlled experiments test a hypothesis by changing one or just a few factors at a time while keeping other factors the same. An *experimental group* is a group of subjects that receives an experimental treatment. A *control group* is a group of subjects that are as similar as possible to the experimental group, with the exception of the experimental treatment. Control groups are used for comparison to test the effect of the experimental treatment.

Randomization and replication are extremely important to good experimental design:

- **Randomization** avoids unintentional selection bias when selecting groups. Individuals should be randomly assigned to groups, or the location of plots should be determined with a random number generator.
- **Replication** increases the reliability of results by lessening the impact of atypical responses. Multiple repeats of each treatment and control should be included in the experiment. Replicates also allow you to do statistical analysis on your results in order to better see patterns in the data. Examples of some statistical analyses are presented later in the chapter.

CONTROLLED EXPERIMENT EXAMPLE

Ecologists wanted to investigate the connections between acorn production, white-footed mouse population, and the risk for Lyme disease. Oak trees produce acorns in a cyclical pattern, making a large crop every two to five years and smaller crops in between. White-footed mice eat acorns, so their populations rise and fall with acorn production. White-footed mice also carry the ticks that host the Lyme disease bacterium.

- The ecologists randomly created six unfenced plots of oak trees in the forest, three experimental plots, and three control plots. Each plot was the same size.
- Mice were initially trapped and removed from all plots at the start of the experiment.
- Four tons of acorns were scattered among the three experimental plots.
- The number of mice and larval ticks was counted on all plots in the year following the addition of acorns.
- More mice and larval ticks were counted on experimental plots than on control plots.
- The scientists concluded that increased acorn production leads to increases in the mouse population, which increases the risk for Lyme disease.

Sampling Methods

Ecologists survey small areas, then extrapolate from their measurements to estimate information about the larger population, for example, to determine population size or species composition. The method of sampling is chosen to fit the population:

» **Random sampling** is used to select an unbiased sample. A random number generator can be used to determine which areas within a larger zone will be sampled to make sure that every part of the area has an equal chance of being chosen. Quadrats, discussed later in this chapter, can be used to mark the areas for sampling.

» **Stratified sampling** takes a proportionate number of observations from each part of the population. For example, if the area to be sampled is 80 percent prairie and 20 percent wooded, then 80 percent of observations would be made in the prairie and 20 percent in the woods.

- **Systematic sampling** is used where the study area includes an environmental gradient, for example, moving up in elevation along a mountain. Transects, discussed later in this chapter, are useful for systematic sampling.

Transects and Quadrats

Transects and quadrats were traditionally used for observing species that are either slow-moving or non-motile, but more recently, transects have also been used for surveying underwater species:

- A *transect* is a line that spans the area of interest. Transects are useful for studying organisms that might have an uneven distribution. Ecologists move along the transect, counting individuals along the way:
  - One method, called the *belt method*, is to count all the organisms of interest within a certain width of the transect line.
  - Another method is to stop and count the organisms in sample plots along the transect, either randomly or at regular intervals.
- To estimate population size using the belt method, divide the total number of individuals counted by the sampled area, then multiply the number by the total habitat area.

TRANSECT EXAMPLE:

You like to scuba, so you are volunteering as a citizen scientist to look for invasive lionfish off the coast of Turkey:

- You count 21 lionfish as you swim along a transect that is 500 meters (m) long and one m wide. The total area sampled is 500 m x 1 m = 500 meters squared (m^2).
- The habitat is 15,000 m^2.
- Estimated population = number counted/area sampled x total area = 21 lionfish/500 m^2 x 15,000 m^2 = 630 lionfish.

QUADRAT EXAMPLE

You are estimating the population of dandelions in a field near your house:

- You randomly scattered ten quadrats measuring 0.5 by 0.5 meters (m) each, which is 0.25 meters squared (m^2).
- The total area of all quadrats is 0.25 m^2 x 10 = 2.5 m^2.
- You counted a total of 62 dandelions among all of your quadrats.
- The total size of your field area is 55 m^2.
- Estimated population size = Quadrat Count/Quadrat Area x Total Area = 62/2.5 x 55 = 1,364 dandelions.

A *quadrat* is essentially a square of a set size used to set up sampling plots in an area. Quadrats are useful for sampling organisms that are distributed evenly in an area. A series of quadrats is placed in a habitat of interest.

» The species within the quadrats are identified and measured, either by being counted or by estimating the percentage of the quadrat occupied by the species.

» Quadrats may be placed along a transect line or scattered randomly in the area of interest.

» To calculate population size from quadrat data, divide the total population count in your quadrats by the total area of your quadrats, then multiply by the total study area.

Mark and Recapture

The Mark-recapture method is a good way to estimate the population size of a motile species.

1. **Catch a predetermined number of individuals at random.**
2. **Mark the individuals and return them to the wild.** Wait to give them time to remix into the population.
3. **Recapture another subset at random.**

MARK AND RECAPTURE EXAMPLE

You want to estimate the number of frogs that live in a pond. You capture ten frogs, mark them, and release them back into the pond. When you revisit the pond in a week, you capture another ten frogs. Two of the frogs you capture are marked.

- The proportion of marked frogs in the recapture is r=2/10=0.2.
- The estimated number of frogs in the pond population is N=10/0.2=50 frogs.

4. **Calculate the proportion of marked to unmarked individuals, represented by *r*.**

   a. Divide the number of marked individuals you recaptured by the total number you captured in the second set.

   b. *r* indicates the fraction of the population your marked individuals represent.

5. **Calculate the estimated number in the population using the formula N=M/*r*.**

   a. The estimated number in the population is represented by N.

   b. M is the number of individuals you originally marked.

Retrospective Studies

Retrospective studies reconstruct ecosystem dynamics and climate change in the past. Ecologists use historical documents and growth markers like tree rings to reconstruct the environment during past centuries. Climate and ecosystem data going back thousands of years can be obtained from sediment cores, ice cores, and archeological research.

RETROSPECTIVE EXAMPLE

Ecologists used ice cores from Antarctica to reconstruct global temperature and atmospheric carbon dioxide stretching back 8,000 years (Chapter 12).

Comparative Studies

Comparative studies compare similarities and differences between different areas, populations, or species. Some examples of possible comparative studies include:

» Comparing the biodiversity (Chapter 13) between two sites that are affected by different environmental factors.

» Comparing the behavior of two different species living in the same habitat.

» Comparing the behavior of one species living in two different habitats.

COMPARATIVE STUDY EXAMPLE

An ecologist compared two different parks located within an urban setting. Two different city organizations managed the sites, and the ecologist wanted to see if the sites were similar enough that they could be considered identical so that management could be transferred to one organization. The ecologist set up their comparative study:

- The ecologist divided each park into grids, then used a random number generator to select three plots in each park for sampling.
- Each plot was sketched and photographed from all corners.
- All plants in each plot were identified, and the space they occupied was estimated.
- Invertebrates were captured from each plot using pitfall traps, which were reset at regular intervals.
- Invertebrates were identified and counted.
- Statistical analysis was used to compare the biodiversity of each park.
- One park had significantly higher biodiversity than the other. In addition, plant species between the sites were different from each other.
- The ecologist concluded that the two parks were significantly different from each other, perhaps due to soil quality or water availability (one site was next to a river).

Lab Experiments

Nature is beautiful, complicated, and messy. Sometimes, in order to tease apart one interaction from all of the ecosystem noise, it helps to bring nature into the lab. In the lab, it's easier to control variables, isolate one or two factors for testing, and set up experiments that have many replicates. Of course, this works best if you're using smaller organisms!

Lab work can also focus on processing samples taken from the field. DNA samples can be extracted and sequenced to identify microbes that live in a particular ecosystem. Pollen can be examined with microscopes as another way to identify plant presence in a sample taken from a living ecosystem or an ancient one. Chemical analysis can reveal plant defenses and soil chemistry. Ecologists today study ecosystems on multiple scales, from the entire system to its smallest molecular components.

ECOLOGY IN THE LAB EXAMPLE

Ecologists wanted to identify the driving force behind predator-prey cycles (Chapter 7) of microscopic animals called rotifers and their prey, green algae. The interactions between these two species didn't follow the classic predator-prey cycles that show a lag between the population growth of predators and prey. Instead, the predator-prey cycles seem to run nearly opposite to each other. The ecologists hypothesized that rapid prey evolution was the most important factor in generating observed patterns. To test their hypothesis, they set up a lab experiment:

- Two different prey populations were cultivated:
  - One prey population contained a single clone, so that all the algae were genetically similar, and rapid evolution by natural selection (Chapter 3) was not likely.
  - One prey population contained seven different clones, so there was a level of genetic diversity in the population.
  - The ecologists added rotifer predators to both prey populations.
  - They tracked the changing population of both predator and prey species.
  - In environments with genetically similar prey, the predator-prey cycles followed the classic pattern rather than the pattern observed by researchers in wild populations.
  - In environments with genetically diverse prey, the predator-prey cycles nearly matched the scientists' original observations. This suggested that genetic diversity was important in the normal cycles, supporting the scientist's hypothesis that rapid evolution of prey species was an important factor.

GIS

Geographic Information Systems (GIS) compile data tied to locations on maps. Ecologists can use GIS to connect maps to information on landscape features like species, vegetation, and habitats. Each type of information is stored on a map layer. Ecologists can use these maps to look at the distribution of certain features over the landscape and look for patterns to see if one factor seems to influence another.

Here are just a few ways that ecologists can use GIS in their research:

- Tracking species migration patterns and habitat changes over time
- Predicting the impact of proposed developments on ecosystems
- Analyzing how urban expansion, agriculture, or logging impacts forest ecosystems
- Determining climate vulnerability zones for droughts, sea level rise, or heatwaves

Modeling

Models are computer simulations of ecological systems. Ecologists program their models with equations that represent relationships among the system's components using information from research studies. Even the most sophisticated computers can't process all of the variables in an ecological system, so computer models are simpler than the real thing.

REMEMBER

Scientists try to *validate* computer models by testing them with real observations. If a model captures the way a system works, then it should be able to accurately predict a known outcome if given the data that led to that outcome.

Models are useful for making predictions about what might happen. For example, climate models allow us to think about possible outcomes of various global warming scenarios. Experiments using established models are cheap and easy to run.

One limitation of models is that they are only as good as the information used to program them. Their simplicity may limit them: If some factors that affect an ecosystem aren't programmed into the model, then real-life outcomes may differ from predictions. Different groups of scientists may develop models for the same system, such as climate change, and these models can make different predictions.

This sometimes confuses people who aren't sure which predictions to believe. Or, people may interpret discussions among scientists about theories or models as serious disagreements about the fundamentals when actually it's colleagues debating the fine details.

Statistics

Ecologists use statistics to analyze their data and look for patterns in the results. They employ many different types of analysis depending on the study design, and often rely on statistical software. Here are some fundamental statistical calculations used in most ecological studies:

- The *mean*, or *average*, is the sum of all numbers, divided by the total number of data points.
- The *median* is the middle number in a data set. If a data set is very variable, the median provides a better indication of the midpoint than the mean because the mean is more affected by high and low data points.
- The *standard deviation* around the mean represents the variability in the data points. Standard deviation conveys the range of data points around the mean, allowing people to see how different two data sets really are. Standard deviation is most easily calculated using spreadsheet software or an online calculator.

Ecologists often use graphs to show their data, plotting the mean values with standard deviation added as error bars around the mean to show the variability. They also use statistical calculations to compare the values between predictions and observations, or between experimental and control plots. These calculations test whether patterns in observed or experimental data are significantly different than predictions based on random chance or the *null hypothesis*, which is the hypothesis that states an experimental variable does not affect the outcome of an experiment.

REMEMBER

Many statistical comparisons generate a value, called the p-value, which indicates whether the patterns in the data are significantly different than the null condition or random chance. The lower the number, the less likely it is that random chance could explain the observations. Scientists usually consider a p-value of less than 0.05 to be strong evidence that the patterns in the data are significant.

STATISTICAL ANALYSIS EXAMPLE

You are studying the impact of gophers on fireweed growth along the slopes of a volcano that recently erupted. You think that, as gophers tunnel, they help distribute seeds in the soil and can speed up recovery. You hypothesize that gophers will increase the growth of fireweed. The null hypothesis is that gophers will not affect fireweed growth.

You have six plots, three experimental and three control. You add a gopher to each of your experimental plots for 24 hours. The experimental plots do not receive a gopher. You come back one week later and count the number of fireweed shoots you see on each plot:

- Your experimental plots (A-C) have 24 (A), 19 (B), and 32 (C) fireweed shoots.
- Your control plots (D-F) have 19 (D), 8 (E), and 7 (F) fireweed shots.

You can do some quick calculations to compare your data:

- » The mean for your experimental plots is: (24+19+32)/3 = 25 plants.
- » The mean for your control plots is: (19+8+7)/3 = 11.3 plants.
- » The median for your experimental plots is 24.
- » The median for your control plots is 8.
- » The standard deviation for your experimental plot is 6.6.
- » The standard deviation for your control plot is 6.7.

It looks like your two plots gave different results. The mean from your experimental plots is larger than that of your control plots. You can estimate whether these numbers are significantly different by considering the standard deviation. Your experimental plot has an average of 25. If you subtract the standard deviation of 6.6, the result is 18.4. The control plot has an average of 11.3. If you add the standard deviation of 6.7, the result is 18. So, even when you consider the standard deviation for each treatment, it looks like your values don't overlap, which suggests that your data is significantly different.

To further test your data, you can do a *t-test*, which is a statistical test that compares two means to see if they are significantly different. You can find calculators online that do the test for you if you put in your data. A t-test of the data from the experimental and control plots confirms that the two results are significantly different, with a p-value of $p = 0.032$.

IN THIS CHAPTER

» Finding our place in the world

» Building resilience through diversity

» Harnessing the power of nature

Chapter 16 Ten Amazing Lessons from Ecology

Earth's ecosystems sustain human populations, leading people throughout history to cast nature as a maternal figure, from Gaia (Ancient Greece) to Sky Woman (Haudenosaunee in North America), to Prithvi (India), and many more around the world. The character of Mother Nature even appeared in margarine commercials of the 1970s. The idea that nature is wise and nurturing persists today in the slogan "Mother (Nature) knows best." This chapter presents ten ways people can apply the lessons of ecology to improve our relationship with our mother.

Everything Is Interconnected

The study of ecosystems (Chapter 10) reveals the intricate ways in which living things are connected to each other and their environment. Humans are part of the interconnected web of ecosystems in which they live. Many human cultures have understood this, promoting the ideal of reciprocity and cautioning against overharvesting from nature. However, some traditions saw humans as dominant over nature, perhaps contributing to the unsustainable exploitation of resources. As human population growth puts increasing stress on the biosphere, people are at risk of losing essential ecosystem services.

REMEMBER

Ecosystem services are the benefits that humans receive from ecosystems.

Ecosystems provide direct and indirect benefits to humans:

- Resources like food, fresh water, building material, and water.
- Regulation of local climate, air, and soil quality, protection against floods and erosion, and storage of carbon.
- Habitat for plants and wildlife, to support the genetic and biological diversity necessary for ecosystem resilience.
- Cultural and spiritual benefits, including opportunities for recreation and relaxation.

More people today are finding creative ways to incorporate ecological thinking into planning decisions, finding ways to conserve local wildlife habitats, restore historic waterways, while also making space for human activity. These efforts represent a shift away from the outdated dichotomy of thinking you have to choose between planning for people or for nature, and towards the reality that planning for people includes planning for nature.

Humans Are Walking, Talking Ecosystems

The human ability to read DNA codes from environmental samples and use them to identify microorganisms has generated new levels of understanding about the types of living things that occupy ecosystems, including the ecosystem of the human body. Scientists estimate that microbial cells in and on the body outnumber human cells by as much as ten to one. Scientists now know more than ever about the identities and possible functions of those microbes, generating a better understanding of the microbial communities that exist in different areas of the body. As scientists refine their understanding of body ecosystems, some wonder what we could learn if we applied concepts learned from the study of the macroscopic to the microscopic.

Here are four lessons from ecosystem science that may provide insight into human health:

- **Biodiversity (Chapter 13) positively influences ecosystem resiliency.** The normal microbiota of the body plays a role in various aspects of human health, from mood to obesity and even cancer. Will greater biodiversity of the human microbiome correlate with better health and resistance to infection? Is biodiversity more important in some areas of the body than others, or during certain periods of the human lifespan?

- **Biological interactions influence community structure (Chapter 8).** The normal microbiota helps resist infection from invading microorganisms by changing the local environment and even releasing compounds that kill other microbes. How do interactions among normal microbes influence communities in the body?
- **Environmental factors influence community structure.** The human body is a diverse landscape with many different microenvironments that vary by food source, pH, salt, and defenses. The communities that inhabit different parts of the body have different structures because of these differences.
- **Community structure changes over time in response to disturbance and succession (Chapter 9).** Babies receive their first dose of microbes from their mother, then their microbes change as they grow, especially in response to disturbances like switching to solid foods. Just like a forest fire can change a landscape, illness and treatment with antibiotics can cause big changes in communities of microbes in the body. What factors influence whether a microbial community recovers after events like these?

Answers to questions like these may change the face of medicine in the future. People already take prebiotics and probiotics in an effort to improve the health of their microbiota. Perhaps someday doctors will view ecosystem maintenance as an important component of preventative health, and treatment options will include prescriptions for microbial therapeutics.

Using Industrial Ecology to Increase Sustainability

Industrial ecology applies ecological principles to product design, considering the life cycle of a product like an ecologist would think about that of an organism. The materials and energy needed to build the product, move the product to market, and dispose of the product at the end of its life are all considered. And just as decomposition recycles the dead, industrial ecologists consider the fate of the materials that were bound up in the product and how best to recycle them. Similar products are considered part of the ecosystem of the product. Considering the entire system of the product helps designers look for levers to maximize sustainable change.

Industrial ecologist Callie Babbitt uses the example of personal computers to illustrate the power of ecological thinking. Even though computers are getting smaller, the environmental impact of computer manufacturing keeps going up, in part because people own more computers and related devices than ever before: computers, smartphones, tablets, printers. . .a single household could own as many as 20 devices. Using ecological thinking, Babbitt points out an opportunity to introduce a single multifunction device as an "invasive species" (Chapter 13) that could replace the need for multiple devices, thus reducing the environmental impact of production. She advocates that, instead of trying to reinvent the wheel, we should look for solutions already crafted by nature over millions of years of trial and error.

The Sun Is a Safe Nuclear Reactor

The sun produces light energy as a product of nuclear fusion reactions between hydrogen atoms in its core. The light travels across space to planet Earth, where photosynthetic organisms capture just one percent of the available energy. One percent is enough to power all the food webs on planet Earth.

During the Industrial Revolution, people discovered they could tap into the remains of ancient photosynthesis in the form of fossil fuels to run their machines, keep lights on in the evening, heat their homes, and travel more quickly. Fossil fuel use shot up, fueling technological innovation. Unfortunately, the consequence of our fossil fuel addiction has now caught up with us in the form of global climate change (Chapter 12).

In an effort to avoid the worst consequences of climate change, people are searching for alternative energy sources. Of the possibilities, climate scientists think solar energy has the greatest potential to reduce carbon dioxide emissions by 2030, the deadline set by the Paris Agreement to cut emissions by 50 percent. In a recent model, scientists calculated that covering half the world's rooftops with solar panels would provide enough energy for everyone on Earth.

Solar farms, areas of land with large solar arrays, can also generate solar power. These arrays represent a disturbance to wildlife and can interfere with animal migration. One way to minimize these impacts is to put solar arrays on land that's already used for agriculture. Some of these *agrovoltaic* projects combine solar panels with grazing animals, others put the panels around beehives.

An even newer approach, called *ecovoltaics*, seeks to produce positive ecological benefits from solar panels. Solar panels typically rotate with the sun, creating changing microclimates of sun, shade, and moisture. Designers of ecovoltaic farms hope to encourage diversity in plant communities around the solar panels.

Trash to Treasure

The study of nutrient cycling in ecosystems (Chapter 11) reveals the extraordinary power of living things to change one form of matter into another. Microbes, in particular, are master recyclers. People have begun to harness this power to clean up their own messes.

REMEMBER

Bioremediation uses natural biological processes to remove toxic environmental contaminants in soil, water, or other environments.

Here are just a few examples of successful bioremediation efforts:

- **Oil-eating bacteria can help clean up pollution after an oil spill.** These bacteria occur naturally in the environment, but in some situations, like oil on a beach, fertilizer is applied to encourage the growth of the bacteria.
- **Microbes can remove heavy metals from soil.** Some microbes can use heavy metal atoms as an energy source, oxidizing them into less toxic materials. Other microbes absorb heavy metals, assimilating them into their cells and removing them from the environment.
- **Plant and microbial communities can clean water.** Small floating rafts of plants can be placed on contaminated water. Microbial communities form on the plant roots, helping to remove contaminants like excess phosphorus that could cause algal blooms (eutrophication, see Chapter 10). The use of plants to clean up contaminants is called *phytoremediation* (Chapter 14).
- **Fungi are effective at breaking down a wide variety of contaminants, including some that normally persist in the environment for a long time.** Fungi have been used to remove polycyclic aromatic hydrocarbons (PAH) from soil. The use of fungi to clean up contaminants is called *mycoremediation*.

Better Together

Living things interact with each other in various ways, from beneficial exchanges like flower nectar for pollen dispersal by a hummingbird, to life-ending moments when a bear eats a salmon. Some of the most fascinating stories are those where two organisms partner in an intimate way, living together in symbiosis (Chapter 7). Some partnerships even give abilities to the collective that neither partner would have on its own.

Lichens, for example, are a mutualistic partnership between a fungus and a photosynthetic organism like an alga. The fungus builds a body that resists water loss, allowing the alga to live on land. The alga performs photosynthesis, sharing some of the carbohydrates with the fungus. The lichen partnership can grow on the surface of rocks, without soil and without access to a steady supply of water.

Some people hope that human communities can increase resilience by taking a lesson from lichens, sharing resources, and supporting one another.

Everything Changes

The history of life on Earth reveals that organisms and even ecosystems change over time. As environmental conditions shift, organisms that are better suited to the new environment are more likely to survive and reproduce, leading to the evolution (Chapter 3) of new species over time. When organisms migrate in response to change, the geographical location of ecosystems may be affected. With increased global temperatures due to climate change, many species are moving towards the cooler temperatures at the poles.

Like other species, humans need to be flexible and ready to adapt their behaviors. This applies to environmental change; for example, planning for ways to help human societies cope with climate change. It can also apply to relationships and business practices. Having a mindset that change is part of life can help people remain open to new possibilities.

Fall Down Nine Times; Get Up Ten

Life on Earth is incredibly resilient. Since life originated 3.5 billion years ago, it has continued in the face of challenges that dramatically altered the environment: ice ages, meteor impacts, fires, volcanic eruptions, and even nuclear explosions. Catastrophic disturbances can reduce an area to bare rock, yet life will return in a process called succession (Chapter 9).

Succession begins with the arrival of pioneer species, first responders who can grow in the harshest conditions, like lichens and some microbes. Pioneer species make subtle changes to the environment, perhaps weathering the rocks, trapping moisture, or creating shade. These changes support the growth of new arrivals, who add their changes. Waves of organisms move into the space, some of the later arrivals crowding or shading out some of the early species. With enough time, you would barely know that a disaster struck.

Many people try to build this same attitude of perseverance and resilience into their own lives. In Japan, there's a phrase, "Nanakorobi Yaoki," which means "fall down seven times, stand up eight." Musician Jon Bon Jovi conveyed the same message when he said, "Success is falling nine times and getting up ten." Perhaps we can all take inspiration from the pioneer species that returned to the slopes of Mt. St. Helens just weeks after it erupted, and that, just 40 years later, most plants and animals have returned.

Strength in Diversity

Biodiversity (Chapter 13) refers to the variety of species in an ecosystem. Having a greater variety of species in an ecosystem increases the likelihood that the system will have some built-in redundancy where more than one species plays a particular role. If a species is lost, another species can fulfill its role, and the system can still function. In an arctic food web, zooplankton might feed on ice algae, phytoplankton, and microbes. If the ice algae were no longer available, the zooplankton could still find food. Zooplankton are food for seabirds, arctic cod, and bowhead whales, so the diversity in their food sources helps support the other species.

In many agricultural systems today, crops are planted in large monocultures where all the plants are genetically very similar. This type of planting is good for success in the short term because the plants are usually very well adapted to the conditions they're planted in. However, these crops are also more susceptible to devastating impacts from diseases or insects: if a pest is capable of killing one of the plants, it can kill them all. And, if the environment changes so that it's no longer favorable for that particular cultivar, then nothing will grow.

We're currently experiencing global climate change (Chapter 12). Global temperatures are rising, and weather patterns are changing. The monocultures of the past may no longer be viable in the future. The lesson from ecosystems is that we need to cultivate diversity to improve our chances of long-term success.

Tread Lightly

Over eight billion people currently live on planet Earth. Just like other organisms, each person draws resources from the biosphere, and each person produces waste. Humans around the world don't make equal demands on the biosphere; however, some people tread lightly while others leave heavy ecological footprints.

REMEMBER

An *ecological footprint* measures the amount of biologically productive land and water required to produce the resources a population consumes and to absorb its waste.

One way to represent a population's ecological footprint is to say how many copies of planet Earth are needed to serve that population. Some populations in Africa need less than one Earth to survive, while some developed nations need between six and nine copies to support their lifestyle.

These calculations raise several concerns:

- **Resources aren't shared equally.** This becomes morally questionable when you consider that some resources, like clean air and water, are global commodities. If one population uses more than its share, others will suffer for it.
- **Some populations are living unsustainably.** *Sustainability* is the concept of living in a way that you maintain resources for future generations. Although some populations are currently able to consume resources at high levels, they are altering the biosphere. For example, populations with large ecological footprints typically burn a lot of fossil fuels. These populations release carbon dioxide to the atmosphere, which is driving climate change and putting stress on the biosphere.
- **The impact of overconsumption is sometimes felt the most by people who use the least.** Some populations in Africa that have small ecological footprints live in areas that are particularly susceptible to drought or don't have high food security. Global warming is changing weather patterns, increasing the risks of drought and threatening global food security. Likewise, some populations in the Pacific who have small ecological footprints live in areas that are at high risk for flooding.

Life is incredibly resilient, but ecosystems can be pushed over tipping points that destabilize the system, leading to major change. Five major mass extinctions (Chapter 13) have already occurred over the history of planet Earth, and many scientists think we are in the midst of the sixth. Hopefully, we can learn lessons from ecology to try to lighten our ecological footprints so we can save the biosphere we know for the many species that live there, including our own.

Index

Numerics

A

B

C

D

E

F

G

H

I

J

M

N

O

P

Q

R

T

U

V

W

X

Y

Z

About the Author

René Fester Kratz, PhD, grew up near the ocean in Rhode Island. As a child, she was inspired by the environmental movement and dreamed of a future with clean air and water for everyone. Her father was a biologist who fed her love of all things science. She graduated from Warwick Veterans Memorial High School and then went on to major in biology at Boston University. At BU, she met and studied under Lynn Margulis, who introduced René to science writing and the world of microbes. René graduated with a bachelor's degree in biology from BU, studied plant ecology at Northern Arizona University, and then went on to get a master's and a doctorate in botany from the University of Washington. At UW, René studied reproductive onset in *Acetabularia acetabulum*, a marine green alga that grows as single cells big enough to pick up with your fingers.

René currently teaches biology and botany at Everett Community College in Everett, Washington. She inspires students to connect with plants and traditional ecological knowledge in her ethnobotany course. In microbiology and global health, she helps students expand their knowledge about the many ways microbes impact human health. René has a strong interest in science education and is passionate about supporting scientific literacy for everyone. She is the author of *Molecular and Cell Biology For Dummies, 2nd Edition; Botany For Dummies, 2nd Edition; Fungi For Dummies;* and *Biology For Dummies, 3rd Edition.*

René loves living in the Pacific Northwest because she is near the ocean and her daffodils start blooming in February. She doesn't mind the rain and thinks the San Juan Islands are one of the most beautiful places on Earth. She visits Rhode Island as often as she can and is grateful for every moment spent with family and friends. René loves traveling to learn first-hand about cultures, history, and plants around the world.

Author's Acknowledgments

Thanks to Matt Wagner of Fresh Books, Inc., for helping me find the opportunity to write this book. And thanks to all the great people at Wiley who made the book happen, especially Elizabeth Stilwell, Rick Kughen, Frederick Dooley, and Kristie Pyles.

Thanks to all my students at Everett Community College for your enthusiasm and hard work. You all inspire me to keep doing what I do.

Dedication

For my sons, Hueston and Dashiel, with the hope that more societies will learn to live in reciprocity with nature, so that we can save the health of our planet.

Publisher's Acknowledgments

Acquisitions Editor: Elizabeth Stillwell
Senior Managing Editor: Kristie Pyles
Development Editor: Rick Kughen
Copy Editor: Rick Kughen
Technical Editor: Frederick Dooley
Production Editor: Tamilmani Varadharaj
Cover Image: Cheryl/stock.adobe.com